UNIVERSITY REVIEWS IN BOTANY

Editor: Professor V. H. Heywood, Ph.D., D.Sc.

3

PLANT PHENOLICS

UNIVERSITY REVIEWS IN BOTANY
Editor: Professor V. H. Heywood, Ph.D., D.Sc.

1. A Dictionary of Microbial Taxonomic Usage
   S. T. Cowan *Handbook*

2. Soil Micro-organisms
   T. R. G. Gray and S. T. Williams

3. Plant Phenolics
   P. Ribéreau-Gayon

In preparation

Biology of the Rhodophyta
P. S. Dixon

Taxonomy and Evolution
V. H. Heywood

Plant Phenolics

PASCAL RIBÉREAU-GAYON
Institut d'Oenologie, Université de Bordeaux II

OLIVER & BOYD
EDINBURGH

OLIVER & BOYD
Tweeddale Court, 14 High Street,
Edinburgh EH1 1YL
A Division of Longman Group Limited

This translation first published 1972

ISBN 0 05 002512 0

This book was originally published by Dunod Editeur, Paris, under the title *Les Composés phénoliques des Végétaux*

Printed in Great Britain by
T. and A. Constable Ltd., Edinburgh

INTRODUCTION

The history of phenolic compounds of plant origin has its beginnings in industry. The longest known class of these compounds, the tannins, have, in effect, been employed since ancient times in tanning skins and the manufacture of ink, and some of their properties have been made use of in the fining of wines. These practices were empirical to begin with, but with the coming of chemistry as a science, it was considered possible that they could be improved by studying their mode of action.

In this manner, therefore, a new chapter of science was written. Chemists distinguished quite clearly various classes of phenolic compounds, some of which were frequently associated with sugars. They included in this family of compounds various pigments such as the flavones and anthocyanins.

Botanists of all persuasions, systematists, cytologists and physiologists, became interested in these substances. They were utilised in taxonomy. Their intracellular occurrence was examined: the observation of cells containing coloured phenolic compounds, especially anthocyanins, gave valuable insights into the origin and evolution of vacuoles. Eventually, studies were undertaken to elucidate the origin of these phenolic compounds and to understand their physiological significance.

This field developed slowly due to the efforts of a small group of researchers, some of whom, moreover, made discoveries of great importance for which their names are well known, as in the case of Willstätter and Karrer. However, some years ago the use of modern research techniques cast doubt upon the greater part of these classical ideas. None the less it is not without value to look back on these concepts since they show that the study of phenolic compounds goes far back into the past.

The present-day situation is superbly presented by Professor Pascal Ribéreau-Gayon in this work which I have pleasure in introducing.

M. Pascal Ribéreau-Gayon belongs to a family of biochemists who for several generations have devoted themselves to the study of oenology. He has, himself, undertaken important research into the chemical composition of the grape and wine, paying special attention to the study of phenolic compounds. He is, therefore, particularly well qualified to write a general work on this subject.

This work is very up to date. Most of the references date from the last 15 years; those which are more than 40 years old are exceptional and refer to works of great historic interest.

M. Ribéreau-Gayon has attempted to bring this book within the reach of non-specialised readers by outlining some of the general principles concerning chemical structure and the chemistry of the phenolics.

The greater part of the book comprises a description of the principal groups of phenolics, their isolation and characterisation. M. Ribéreau-Gayon has clearly attempted to provide the research worker with a tool which can be used in a wide variety of circumstances. The techniques of chromatographic isolation and spectrographic analysis are presented in careful detail. Moreover, many of these are original.

The biosynthesis of phenolic compounds has not been neglected. On reading the chapter devoted to this subject, the reader will note that the only valid results were obtained by the method of ^{14}C labelling. But these results so far have scarcely had any physiological bearing. The role of phenolic substances remains very uncertain. Do they take part in the formation of growth substances? Do they contribute to the phenomena of oxidation-reduction? Are they connected with photosynthesis? Or do they have no essential function at all? At present we do not know.

Let us hope that substantial future efforts of chemists will concentrate on this question of physiology. And if this development should take place, we can be certain that it will be due in large measure to this work of M. Ribéreau-Gayon. This book is most timely. It will be of great value and we are certain that it will enjoy a great success.

R.-J. GAUTHERET
Membre de l'Institut

PREFACE TO THE FIRST EDITION

Plant tissues contain a large number of substances with phenolic functions, the most important of which are the anthocyanins (red and blue pigments of flowers and fruits), the flavones and their derivatives (yellow pigments) and, lastly, the tannins; one can add to these the phenolic acids whose wide distribution in plants has recently been recognised.

The study of phenolic compounds has benefited particularly from the contribution of paper chromatography, a technique which has played an important role in the recent development of many different branches of biochemistry.

Without doubt the classical methods of chemistry permitted the elucidation, long ago, of the structure of the main substances in this family of compounds. In the field of anthocyanins, the fundamental work of R. Willstätter, P. Karrer and R. Robinson remains justly celebrated; moreover the Nobel Prize in chemistry was awarded to these three *savants* in 1915, 1937 and 1947 respectively for their research into natural compounds. But these purely chemical methods are difficult to apply; they do not lend themselves well to the separation and identification of all the constituents in complex mixtures which are found frequently in plant tissues; they have not allowed a complete inventory of phenolic compounds of plants to be carried out.

On the other hand, paper chromatography, whose operation is more delicate than we might suppose and which always ought to be used in conjunction with chemical methods, lends itself well to the separation of all the phenolic compounds of a plant organ and also to their identification, since a chromatographic R_f value has as much value as a melting point as a criterion of purity and identification. Since the first publication of Bate-Smith on this subject in 1948, many studies have been carried out; these have made an important contribution to this chapter of plant biochemistry for which we have detailed information available only in recent years.

This has recently led to the publication of several review volumes, of which the two most important are those of Geissman (1962) and

Harborne (1964). The work of Sannié and Sauvain is the only one available in French, but it deals only with flavonoids and does not cover the phenolic acids or the tannins. On the other hand, it was published in 1952 and did not, therefore, include the many contributions made by paper chromatography to this field.

I have been stimulated to write a book on phenolic compounds because of this absence of reviews in French and also, or course, because of my own personal research interest in these substances since 1953. After learning chromatographic techniques, which were then quite new, in the biological chemistry section of Professor M. Macheboeuf at the Institut Pasteur, Paris, I spent long periods in the biochemical laboratory of Dr E. C. Bate-Smith at Cambridge and in the organic chemical laboratory of Professor T. A. Geissman at the University of California, Los Angeles.

To undertake a survey of the phenolic compounds in a plant tissue today has become a much simpler operation than ten to fifteen years ago; it is not, however, always a simple matter. I have sought in this work to describe the different operations involved in carrying out these identifications so as to make them available to relatively unskilled workers. Since phenolic compounds possess many important properties, their study is becoming more and more useful in more general research in physiology, biology and even botany, as well as in food technology; such research is often undertaken by workers who do not have a detailed chemical training. For this reason I have tried to describe in detail the various reactions involved in the different analytical operations and to interpret them in terms of the properties of the benzene ring and phenolic function; a special chapter has been devoted to this question and I have not hesitated to refer to basic ideas of organic chemistry which may seem elementary to some readers.

In the various scientific disciplines mentioned above which call on phenolic compounds, numerous studies have been carried out and important results obtained, but some developments are still necessary. Further, without underestimating the importance of these problems I have accorded them a less important place in this treatment than that devoted to analytical problems. All these matters are discussed in the last chapter, as well as the biosynthesis of phenolic compounds which has been widely studied and whose basic concepts have now been established.

It is because of a particular application of the knowledge of phenolic compounds that I became interested in them. When attached to an organisation specialising in the study of the theoretical and practical problems posed by oenology (the science of wine) some fifteen years ago, I was struck by how little attention was paid to the role of phenolics in the traditional techniques involved in the making and preservation of wine, certainly out of proportion to the real value of these compounds (they are responsible, particularly, for all the differences between white and red wines). As this situation was evidently due to our lack of basic knowledge, I was obliged first of all to obtain the necessary basic information for solving the practical problems. It is these general notions, and more particularly the analytical methods which I have had occasion to employ, that are described in this book; however, I shall refer frequently, by way of example, to the particular case of the grape and wine.

I should like to express my sincere gratitude to Professor R.-J. Gautheret, Membre de l'Institut, for kindly agreeing to write an introduction to this book. Biochemist by training, I am particularly happy that a biologist of authority, especially in the field of tissue culture, should have indicated in this way his interest in the use of chemical methods in the study of phenolic compounds.

I would also like to thank Professor Jean Ribéreau-Gayon, whose advice and support has always encouraged me in my studies and more particularly in the preparation of this work. In the Services d'Oenologie of the University of Bordeaux of which he is director, M. J. Ribéreau-Gayon has always aimed, not without difficulty, to coordinate fundamental and applied research activities for the greater effectiveness of both; I should like this book, and the research on which it is based, to be a testimony to the success, both scientific and technical, of this undertaking.

P. R.-G.

Bordeaux,
October 1967.

PREFACE TO THE ENGLISH EDITION

Because of the increasing significance of phenolic compounds in many fields of botanical and biochemical research, the need for a concise account of their chemical nature, methods of identification, biosynthesis and taxonomic and physiological importance has often been voiced. Professor P. Ribéreau-Gayon's book *Les composés phénoliques des végétaux* admirably filled this gap in the literature, and it was at the suggestion of Dr E. C. Bate-Smith that it was eventually decided to prepare an English translation for inclusion in the *University Reviews in Botany* series.

Professor Ribéreau-Gayon has kindly taken the opportunity of this translation to revise and bring his work right up to date so that the English version constitutes a substantial revision of the original French text.

It is a felicitous circumstance that most of the translation has been prepared by three scientists who themselves occupy a very distinguished position in the field of plant phenolics—Dr E. C. Bate-Smith (chapters 1-5), Dr J. B. Harborne (chapter 6), Dr T. Swain (chapter 7). Mr I. B. K. Richardson translated chapter 8.

My colleague, Dr Harborne, has played a major rôle in the scientific preparation and collation of the manuscript and has greatly assisted me in editing the translation. I am very grateful to him for his support.

V. H. Heywood
Editor, University Reviews in Botany

July 1971

CONTENTS

1

Conspectus of the Phenolic Constituents

Introduction

It is not the intention of this book to list every plant constituent that possesses a phenolic hydroxyl function. These are far too numerous, and it is proposed, in Chapters 4-7, to deal only with those which are widely distributed in the plant kingdom. It is, however, necessary to give, in this first chapter, a classification of the phenolic constituents which are actually known at the present time.

Harborne and Simmonds (1964) have provided such a classification:

C_6 :	Simple Phenols
C_6—C_1 :	Phenolic Acids and Related Compounds
C_6—C_2 :	Acetophenones and Phenylacetic Acids
C_6—C_3 :	Cinnamic Acids and Related Compounds
C_6—C_3 :	Coumarins, Isocoumarins and Chromones
C_{15} :	Flavones
C_{15} :	Flavanones
C_{15} :	Isoflavones and Isoflavanoids
C_{15} :	Flavonols, Dihydroflavonols and Related Compounds
C_{15} :	Anthocyanidins
C_{15} :	Chalcones, Aurones and Dihydrochalcones
C_{30} :	Biflavonyls
C_6—C_1—C_6, C_6—C_2—C_6 :	Benzophenones, Xanthones and Stilbenes
C_6, C_{10} and C_{14} :	Quinones
C_{18} :	Betacyanins

The principal phenolic constituents are not present in a free state in nature, but in the form of esters or, more generally, as glycosides. It is convenient, however, to include certain polymers such as lignin and tannins, which are important natural products formed by the condensation of molecular units belonging to one of the above listed groups.

In a recent review, Swain and Bate-Smith (1962) distinguished between 'common' and 'less common' phenolic constituents. Such a distinction appears to be essential. Anyone approaching this subject for the first time is in danger of being put off by the complexity of the structures that he will encounter, and will have great difficulty in singling out, from the multitude, those constituents that merit deeper consideration.

The treatment adopted here has been determined by the need to select what is important, without losing sight of essential facts or evading intrinsic complexity. In order to do this, the substances have been assembled into three groups:

The families of widely distributed phenolic constituents.

The families of less widely distributed phenolic constituents.

Phenolic constituents present in nature in polymeric form.

Generally speaking, a family of 'less widely distributed phenolic constituents' implies a family in which a limited number of particular members is known, each of which has only very rarely been identified. Conversely, in the case of families of 'widely distributed phenolic constituents' some substances are virtually universally distributed in the plant kingdom, while others may have been identified in only one plant, or at any rate in only a very few species. In the chapters of this book devoted to a detailed study of particular families of widely distributed phenolic constituents, consideration will be given in each instance only to those substances of especial importance.

This first chapter brings together the different families of phenolic constituents, with examples of the structures of particular individual substances. But each family contains many substances, varying in the nature of the substituents (OH or OCH_3) carried by the aromatic skeleton that they all possess in common.

Further details of the phenolic constituents and their distribution, especially of those which are less widespread, are to be found in the reviews of Geissman and Hinreiner (1952), Bate-Smith (1962b) and Harborne and Simmonds (1964).

The different families of widely distributed phenolic constituents

1.1 *General structure of the flavonoids*

These constituents are classified in four groups as follows:

Benzoic acids, cinnamic acids and coumarins.

Flavones, flavonols and related compounds.

Chalcones, dihydrochalcones and aurones.

Anthocyanins.

This method of presentation is conventional; for instance, while it is clear that there is no obvious relationship between the benzoic acids and the coumarins, there is, however, one between the coumarins and the cinnamic acids.

The substances placed in the last three groups constitute the 'flavonoids', characterised by having the C_6—C_3—C_6 structure in common, in which two benzene rings are linked by a C_3 group which is different in the different flavonoids.

In the great majority of cases, the A ring (the left-hand ring) is either *meta*-dihydroxylated (resorcinol type) or *meta*-trihydroxylated (phloroglucinol type); except for the chalcones (e.g. butein, 2), one of the hydroxyls of this A ring is combined in an oxygen heterocycle of five (4) or six (1, 3, 5) atoms. By contrast, the B ring is either monohydroxylated (3), *ortho*-dihydroxylated (1, 2, 4) or *vic*-trihydroxylated (5). This difference is due to the fact that the two rings have different biosynthetic origins: ring A is formed by the condensation of three molecules of acetic acid (8.3), whilst the B ring is derived from sugars by the shikimic route (8.2). In addition, the different OH groups may be methylated.

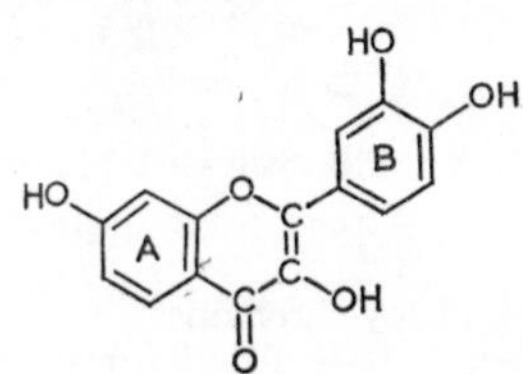

(1) FLAVONOL: fisetin

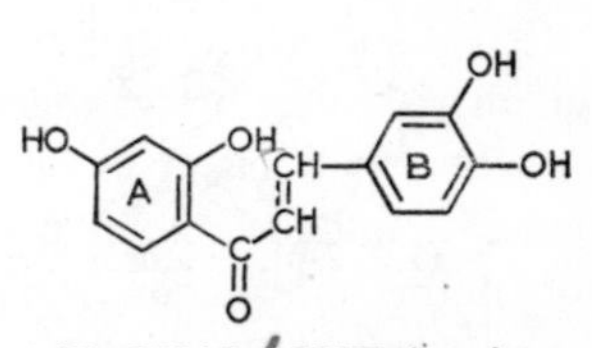

(2) CHALCONE: butein

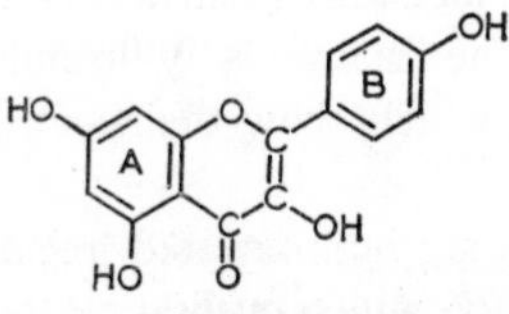

(3) FLAVONOL: kaempferol

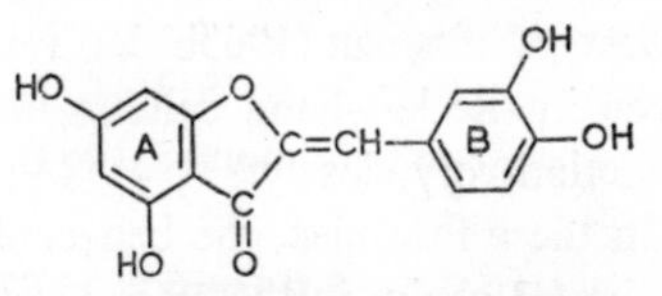

(4) AURONE: aureusidin

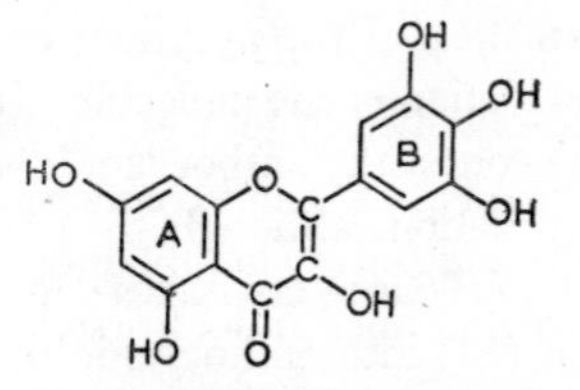

(5) FLAVONOL: myricetin

The structure of the different types of flavonoids differs according to the nature of the oxygen heterocyle; this heterocyle derives either from pyran (6), pyrylium (7) or γ-pyrone (8). Except in the case of the aurones (e.g. aureusidin, 4), cyclisation takes place between the third carbon of the chain and an OH group of the A ring *ortho* to this chain, resulting in chroman (9), chromene (10) or chromone (11) structures, such as 2-phenylchroman (12) or 2-phenylchromone (13).

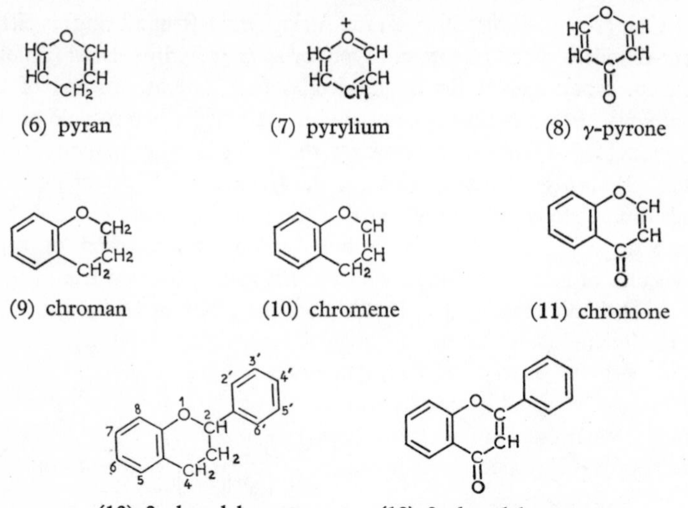

The C3 residue varies in its level of oxidation, that of the flavonols being the highest and that of the catechins and dihydrochalcones the lowest. Geissman (1963b) has represented the flavonoids by formulae, given in the left-hand column of Table 1, which show their state of oxidation very clearly.

In these formulae, the benzenoid A rings are hydroxylated, relative to the C3 chain, in the *ortho* position, through which cyclisation takes place, in the course of which such unique reactions as enolisation, elimination of water, or formation of the pyrylium ion takes place without change in the state of oxidation of the molecule. These formulae show, for instance, that the flavones, the anthocyanidins and the aurones are at the same level of oxidation, although they possess such different structures. Geissman (1963b) envisages the transformation of A—CH_2—CO—CO—B (14) into anthocyanidin (17) through a route involving cyclisation (15) followed by enolisation to a structure corresponding with the pseudobase of the anthocyanidins (16) which in acid solution, gives rise to the oxonium form of the anthocyanidin (17) (6.4). Geissman (1963b) also draws attention to the fact that the minimum state of oxidation (18), corresponding with flavonoids of the structure (19), are unknown in nature.

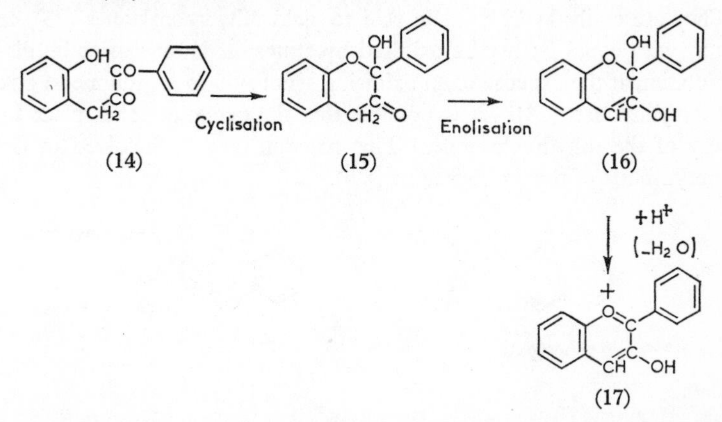

(14) (15) (16) (17)

It is necessary to add to the flavonoids in Table 1 the isoflavones (20) (and probably also the isoflavanones), compounds which are isomeric with the C_6—C_3—C_6 structures so far described. Substances are also known, the neoflavonoids (Lund 1968a) such as dalbergin (21), in which the B ring is attached to carbon atom 3 of the central chain.

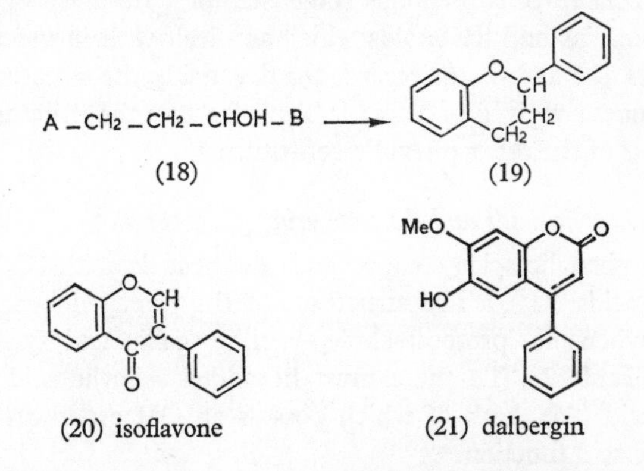

(18) (19)

(20) isoflavone (21) dalbergin

The flavenes constitute a family of flavonoids which do not exist as such in nature but appear through the transformation of a natural product (6.5, 7.8). They are characterised by the absence of a CO group and the presence of a double bond in the central heterocycle, either between carbon atoms 2 and 3 (flav-2-ene) or between carbon atoms 3 and 4 (flav-3-ene). These substances have been especially studied by Lund (1966) and Weiss and Lund (1968).

The flavenes (22) are obtained by reduction of the corresponding flavones (R=H) or flavonols (R=OH) with lithium aluminium hydride. They are easily hydrolysed in acid to yield dihydrochalcones (23), or may with equal facility be oxidised by atmospheric oxygen or by dismutation at the expense of an oxidising agent such as a quinone to give a flavylium salt (24). It is possible that the oxidation takes place by way of the dihydrochalcone. This reaction may be involved in the biosynthesis of the anthocyanins (8.5).

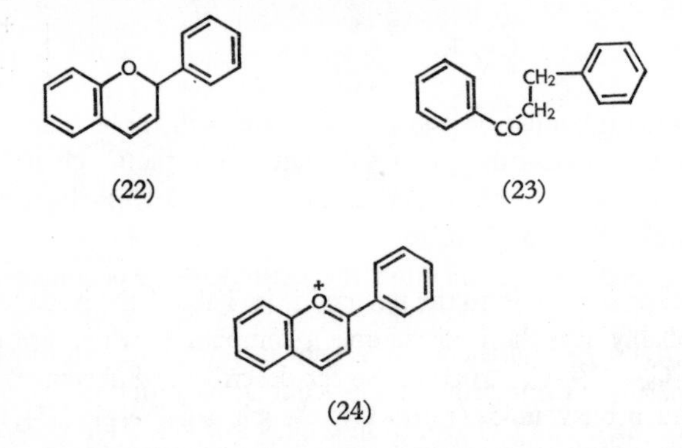

In the next three subsections some examples are given of phenolic acids, coumarins and flavonoids which are dealt with in greater detail in chapters 4, 5 and 6. As regards the flavonoids, the structures of the principle ones are given in Table 1: Table 2 shows the structures of the commonest of the other phenolic constituents.

1.2. *The phenolic acids and the coumarins* (Chapter 4)

The term phenolic acids comprehends the benzoic acids (C_7) and the cinnamic acids (C_9). The structures of the principal benzoic acids (*p*-hydroxybenzoic, protocatechuic, vanillic, gallic and syringic) are given in Table 2. To these must be added salicylic acid (25) and gentisic acid (26), both of which possess an OH group *ortho* to the carboxylic acid function.

The principal cinnamic acids are four in number (Table 2). Practically all plant tissues contain at least one of them. *p*-Coumaric acid and caffeic acid are among the commonest of all the phenolic constituents; ferulic and sinapic acids are also widely distributed.

All the phenolic acids occur naturally in combination, usually in the form of esters. These combinations are actually well known in the case of the cinnamic acids (combinations with quinic acid, sugars, tartaric acid); the ester of caffeic acid with quinic acid, chlorogenic acid, is the classic example. The case of the benzoic acids is quite different, so little being known of the forms of combination in which they occur.

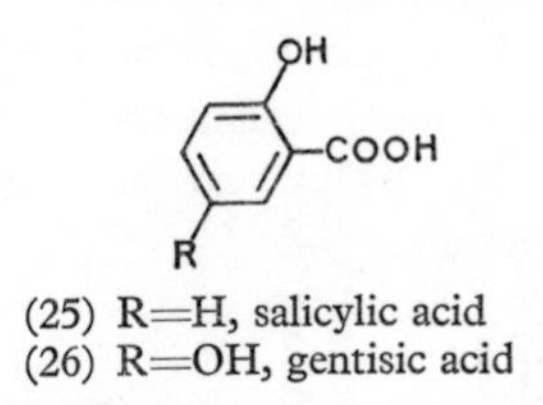

(25) R=H, salicylic acid
(26) R=OH, gentisic acid

The coumarins must be considered along with the cinnamic acids, since they also possess the C_6—C_3 configuration, the C_3 chain in their case being in the form of an oxygen heterocycle. Umbelliferone (27) is an example of a coumarin.

It is also necessary to mention the existence of isocoumarins (e.g. bergenin, 28); these are phenolic constituents with a more restricted distribution, only six of them being known compared with upwards of fifty coumarins. Coumarins and isocoumarins differ in the relative positions of the O and CO groups in the heterocycle. They are chemically related to the chromones (47) (1.8).

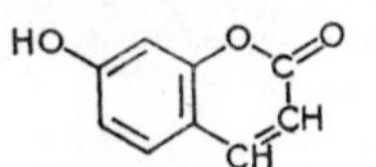

(27) COUMARIN: umbelliferone

(28) ISOCOUMARIN: bergenin

1.3 *The flavones, flavonols and related substances* (Chapter 5)

Of these compounds, the flavonols (3-hydroxyflavones) are by far the most widely distributed. The structures of the three commonest flavonols, quercetin, kaempferol and myricetin, are given in Table 2. Quercetin is without any question the phenolic compound with the widest distribution in nature. The flavones *sensu stricto* are functionally of less importance that the flavonols. Apigenin and luteolin, which have hydroxylation patterns corresponding with kaempferol and quercetin respectively, occur however quite frequently in many families of

Table 1. *The different flavonoid classes*

Structure of the central C_3 unit	Class	Principal substances: Name	Principal substances: Hydroxylation (*)
A—CH_3—CHOH—CHOH—B →	Catechins (flavan-3-ols)	catechin gallocatechin	5,7,3′,4′ 5,7,3′,4′,5′
A—CO—CH_2—CH_2—B →	Dihydrochalcones	phloretin hydroxyphloretin	3,2′,4′,6′ 3,4,2′,4′,6′
A—CO—CH=CH—B →	Chalcones	butein	3,4,2′,4′
A—CO—CH_2—CHOH—B →	Flavanones (dihydroflavones)	naringenin butin eriodictyol	5,7,4′ 7,3,′4′ 5,7,3′,4′
A—CHOH—CHOH—CHOH—B →	Leucoanthocyanidins (flavan-3,4-diols)	leucocyanidin leucodelphinidin	5,7,3′,4′ 5,7,3′,4′,5′

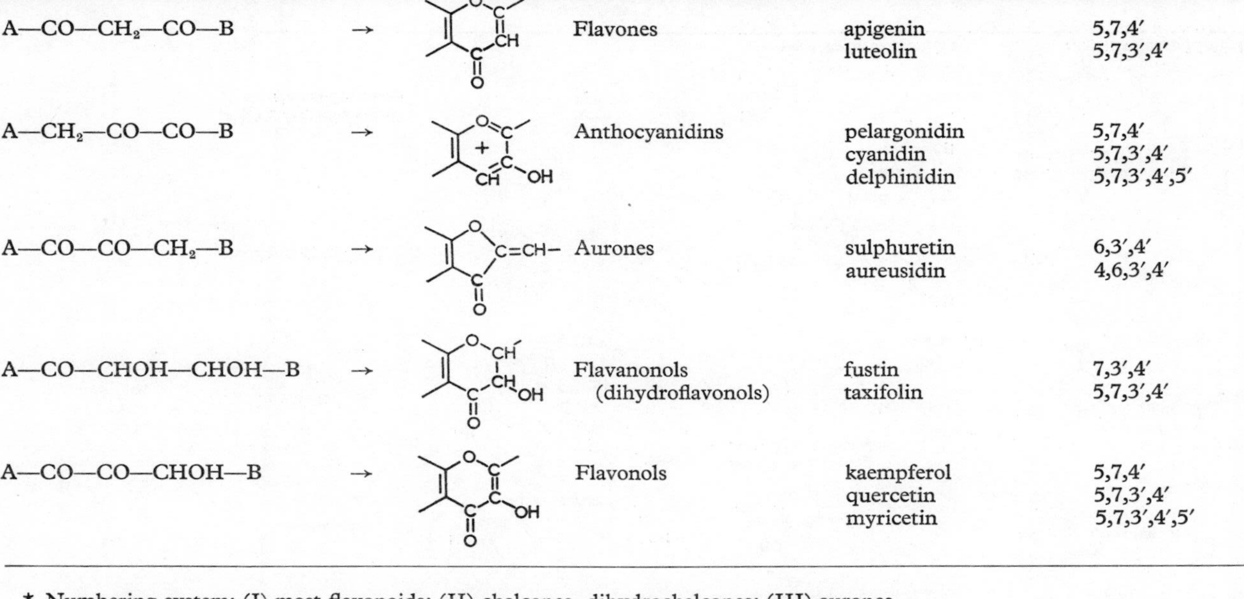

A—CO—CH_2—CO—B	→	Flavones	apigenin luteolin	5,7,4′ 5,7,3′,4′
A—CH_2—CO—CO—B	→	Anthocyanidins	pelargonidin cyanidin delphinidin	5,7,4′ 5,7,3′,4′ 5,7,3′,4′,5′
A—CO—CO—CH_2—B	→	Aurones	sulphuretin aureusidin	6,3′,4′ 4,6,3′,4′
A—CO—CHOH—CHOH—B	→	Flavanonols (dihydroflavonols)	fustin taxifolin	7,3′,4′ 5,7,3′,4′
A—CO—CO—CHOH—B	→	Flavonols	kaempferol quercetin myricetin	5,7,4′ 5,7,3′,4′ 5,7,3′,4′,5′

* Numbering system: (I) most flavonoids; (II) chalcones, dihydrochalcones; (III) aurones.

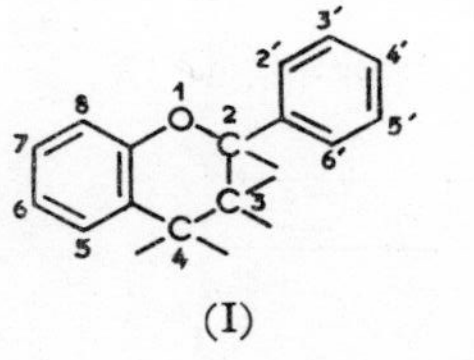

(I)

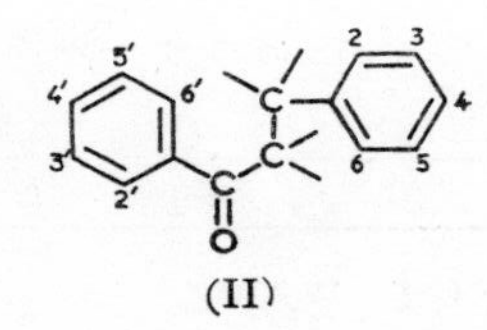

(II)

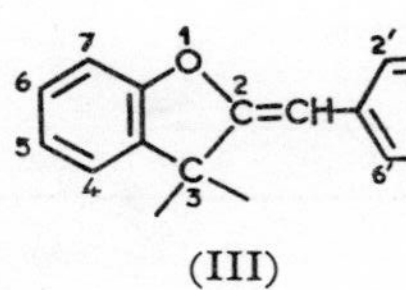

(III)

Table 2. *The major natural phenolics*

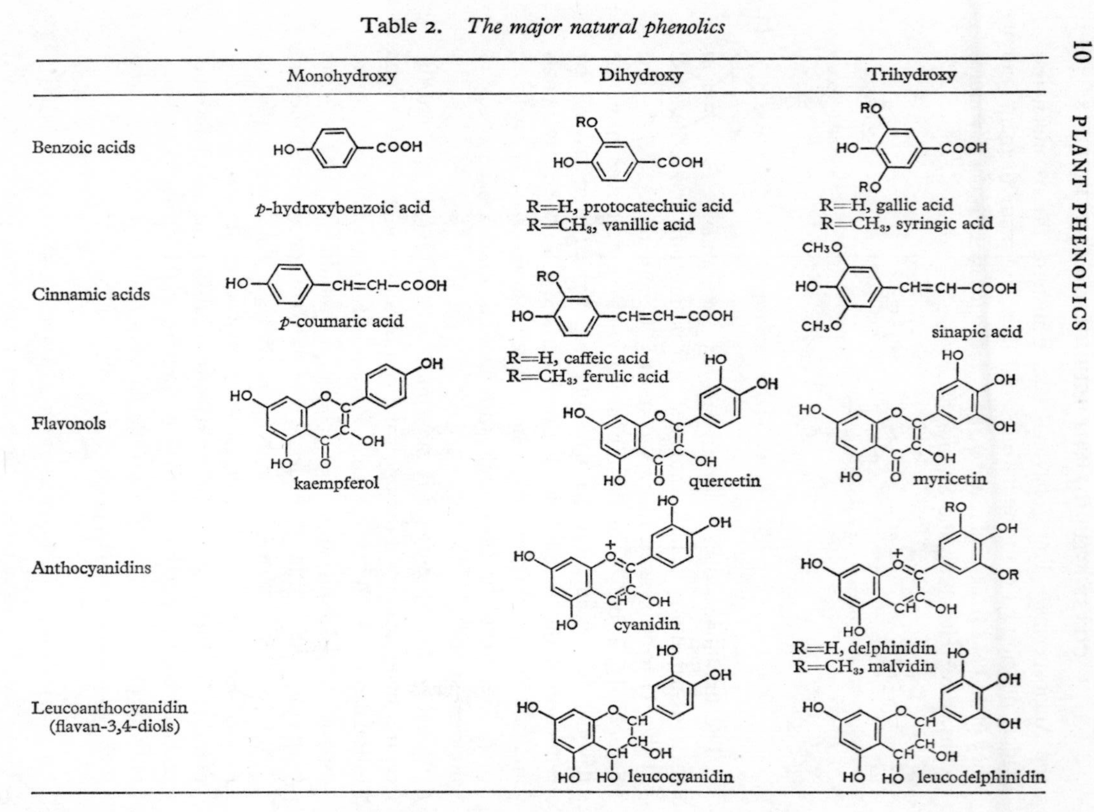

	Monohydroxy	Dihydroxy	Trihydroxy
Benzoic acids	*p*-hydroxybenzoic acid	R=H, protocatechuic acid R=CH_3, vanillic acid	R=H, gallic acid R=CH_3, syringic acid
Cinnamic acids	*p*-coumaric acid	R=H, caffeic acid R=CH_3, ferulic acid	sinapic acid
Flavonols	kaempferol	quercetin	myricetin
Anthocyanidins		cyanidin	R=H, delphinidin R=CH_3, malvidin
Leucoanthocyanidin (flavan-3,4-diols)		leucocyanidin	leucodelphinidin

the Angiosperms. The isoflavones (e.g. genistein, 29) are much less widely distributed than the flavones, but there are many of them, and they often have structural features rarely met with in the other flavonoids (Ollis, 1962; Harborne and Simmonds, 1964).

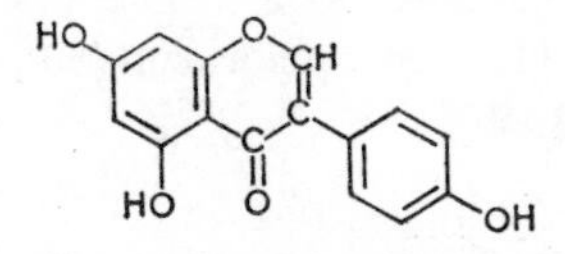

(29) ISOFLAVONE: genistein

The flavanones, or 2,3-dihydroflavones, are similarly restricted in their distribution. The commonest are naringenin and eriodictyol (Table 1) hydroxylated like apigenin and luteolin respectively. The flavanonols (Table 1), or 2,3-dihydroflavonols, are derived from the flavonols in the same way that flavanones are derived from flavones, by reduction of the double bond between C2 and C3. The flavanonols are, in a word, 3-hydroxyflavanones. Such substances (e.g. taxifolin) occur essentially in association with tannins in the heartwoods of different species.

Finally, the existence of flavans must be mentioned; in these the central heterocyle is on the one hand completely saturated and on the other hand has no CO group. Flavan-3-ols (catechins) and, more especially, flavan-3,4-diols (or leucoanthocyanidins) (Table 1) are frequently found in plant tissues, where they are concerned in the formation of the condensed tannins (Chapter 7). The most important flavans are catechin and gallocatechin, leucocyanidin and leucodelphinidin (Tables 1 and 2). The flavans are distinguished from other phenolic constituents in that they are present in the form of aglycones, most frequently polymerised, while flavones, flavonols and their relatives are always* present in glycosidic combination (1.6).

1.4 *The chalcones, dihydrochalcones and aurones* (Chapter 5)

These are all phenolic compounds having the structure C_6—C_3—C_6, but the C_3 element is either not heterocyclic, or is in the form of a pentacyclic ring (aurones). Chalcones and aurones are the yellow pigments of some flowers, their colour changing to red or orange-red in presence of ammonia. The numbers of these substances known are

* There are, however, occasional exceptions in the form of exudates from the surfaces of leaves and secretions into internal cavities. (Translator's note—E. C. B-S.)

relatively few and they are limited in their distribution. As examples, the chalcones are represented in Table 1 by butein and the aurones by aureusidin and sulphuretin.

Not very much is known about the dihydrochalcones, derived by hydrogenation of the double bond. The best known is phloretin (Table 1), present as a glucoside, phloridzin (30), in various parts of the apple tree (Williams, 1960).

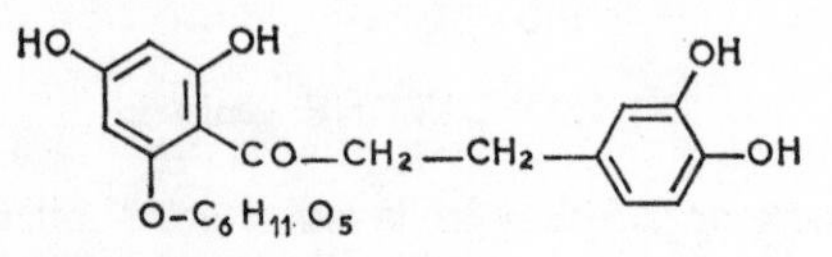

(30) phloridzin (phloretin 2′-glucoside)

1.5. *The anthocyanins* (Chapter 6)

These are red pigments in acid solution, turning blue in alkali and in certain other circumstances, such as when they form complexes with metals. They are very widely distributed in flowers and fruits. They have a C_6—C_3—C_6 structure in which the C_3 element is in the form of a pyrylium ion, and they are cations. These are, without question, the most interesting of all the phenolic compounds. They occur naturally as glycosides, very many of which are known, but the aglycones from which they are derived are quite few in number, only six being at all widely distributed. Table 2 gives the structures of cyanidin, delphinidin and malvidin, the first being by far the most common.

1.6. *Combined forms of the flavonoids*

It is now well known that the phenolic constituents normally exist in plants in combined form and, as has already been mentioned (1.2), this form is, in the case of the phenolic acids, usually that of an ester. In the case of the flavonoids, the most frequent form of combination is glycosidic, in which the bond —C—O—C— is formed between an alcoholic or phenolic hydroxyl of the phenolic compound and an OH group of a sugar molecule (2.5).

While, theoretically, the glycosidic link might be formed with any one of the hydroxyl groups of the flavonoid molecule, in actual practice it is found that certain positions are favoured. For instance, the anthocyanins are nearly always* glycosylated at position 3, and frequently

* With the necessary exceptions of the 3-desoxyanthocyanidins, apigeninidin and luteolinidin, and of some 7-glycosides reported in a few plants. (Translator's note—E. C. B-S.)

also at position 5. In the case of the flavones, the most favoured position is 7, and in the flavonols, 3 (the numbering is given in Table 1). Of the sugars found in glycosidic combination, D-glucose is by far the most common, but L-rhamnose, D-galactose, L-arabinose, D-xylose and several di- or trisaccharides, e.g. rutinose, may be found. For every aglycone, therefore, there may be many different glycosidic combinations, which is the reason for the very large number of phenolic compounds known to occur in the plant kingdom.

In any one organ there may be many constituents difficult to separate and identify. When, therefore, one studies unfamiliar plant material, it is always best to begin by studying the simple phenolic substances liberated by hydrolysis of their glycosides or esters, before embarking on the analysis of their numerous structural combinations. Only in the case of the flavans (catechins and leucoanthocyanidins) are the flavonoid residues uncombined with sugar, but these are usually polymerised to form the condensed tannins.

The different families of less widely distributed phenolic compounds

1.7. *The* C_6, C_6—C_1 *and* C_6—C_2 *phenolic constituents*

The simplest phenols, phenol itself (31), catechol (32), hydroquinone (quinol) (33), resorcinol (34) and phloroglucinol (35) have on occasion (except for resorcinol) been reported as occurring in nature, but they are not frequently encountered. However, catechol, phloroglucinol and, occasionally, resorcinol residues occur in the more complicated molecules of the flavonoids (1.1).

The aldehydes are included in this family with equal justification; they are derivatives of the benzoic acids, which can most often be

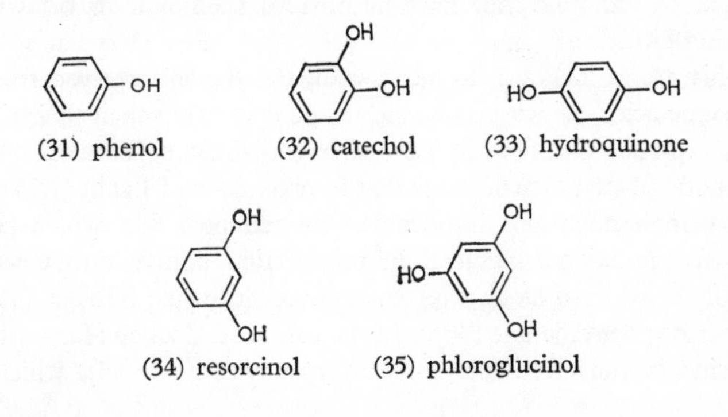

(31) phenol (32) catechol (33) hydroquinone

(34) resorcinol (35) phloroglucinol

regarded as constituents of the essential oils. *p*-Hydroxybenzaldehyde (36), vanillin (37) and syringaldehyde (38), or at any rate substances yielding these aldehydes under alkaline conditions, have been reported (Pearl, 1958) as constituents of the wood of conifers. They are closely related to the structure of lignin.

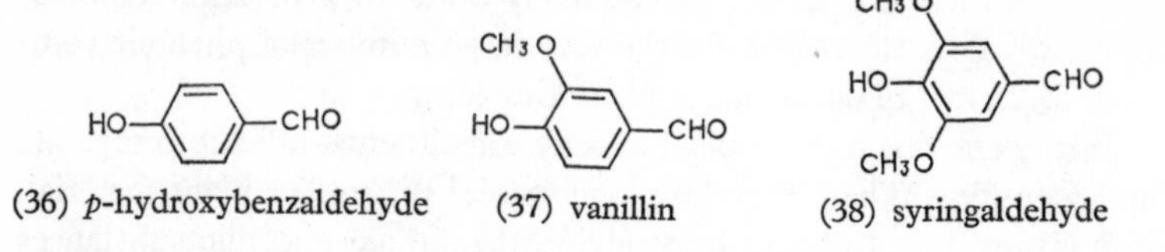

(36) *p*-hydroxybenzaldehyde (37) vanillin (38) syringaldehyde

The alcohols derived from benzoic acids (C_6—C_1) are relatively uncommon, but salicyl and gentisyl alcohols are occasionally found; 3,4-dihydroxybenzyl alcohol has recently been identified (Challice and Williams, 1968a) in *Pyrus calleryana*, in the form of a glycoside acylated with a phenolic acid. The phenolic compounds with a C_6—C_2 skeleton are among the least frequent. Compounds derived from acetophenone (2-hydroxyacetophenone, 39) or phenylacetic acid (2-hydroxyphenylacetic acid, 40) have occasionally been isolated.

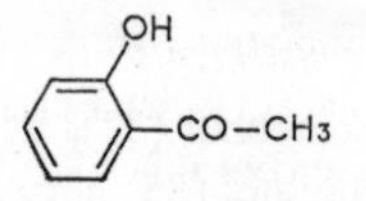

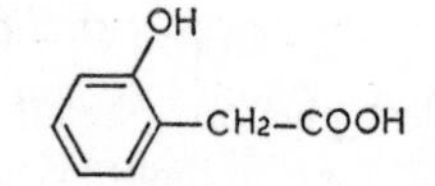

(39) 2-hydroxyacetophenone (40) 2-hydroxyphenylacetic acid

1.8. *The* C_6—C_3 *phenolic constituents*

The C_6—C_3 skeleton is unquestionably the commonest and most important. A large number of phenolic compounds possess this structure. They are dealt with collectively, in this paragraph, for the sake of simplicity, although they have no obvious taxonomic or biosynthetic relationships.

This group includes, to begin with, the alcohols derived from the cinnamic acids, which occur much more commonly than their C_6—C_1 counterparts. Coniferyl (41) and sinapyl (42) alcohols are constituents of woody plants; they are regarded as precursors of lignin (1.15).

Attention must also be given to two amino acids, which possess phenolic functions, included in this section although they are not usually considered along with phenolic constituents: tyrosine (43) and dihydroxyphenylalanine (44). In the course of alcoholic fermentation, tyrosine is converted into *p*-hydroxyphenylethanol (45), which is a normal constituent of fermented beverages, beer (McFarlane and Thompson, 1964) and wine (Ribéreau-Gayon and Sapis, 1965). To these structures may also be added various aromatic amines, such as tyramine (46), which are sometimes classified as alkaloids.

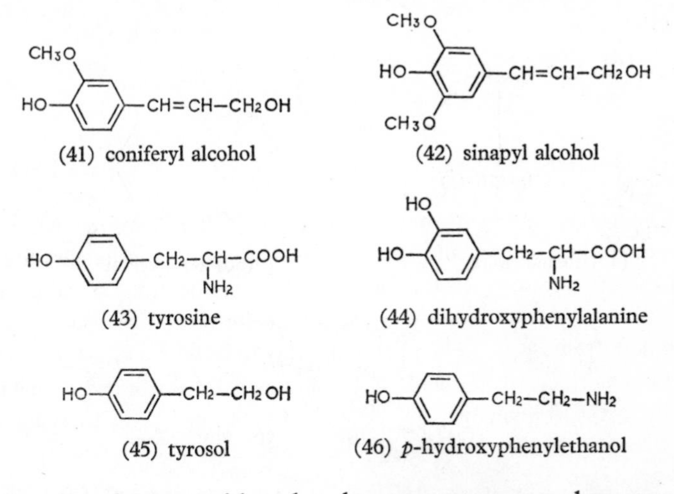

(41) coniferyl alcohol (42) sinapyl alcohol

(43) tyrosine (44) dihydroxyphenylalanine

(45) tyrosol (46) *p*-hydroxyphenylethanol

Phenolic substances with a phenylpropane structure such as eugenol (47) and isoeugenol (48) should also be placed in this class of C_6—C_3 compounds. These are constituents of essential oils.

Here also are to be included the lignans, C_6—C_3 dimers which are heartwood constituents associated with lignin. A number of them have been described by Hathway (1962). They are of different structural types, such as the dibenzylbutyrolactones (e.g. matairesinol, 49) or the diphenyltetrahydrofurofurans (pinoresinol, 50).

The last class of C_6—C_3 compounds to be mentioned are the chromones, which are related to the coumarins, although much less common. The simplest example is that of eugenin (51).

1.9. *The* C_6—C_1—C_6 *and* C_6—C_2—C_6 *phenolic constituents*

Under this heading are grouped, following Harborne and Simmonds (1964), the relatively rare derivatives of benzophenone (e.g. maclurin (52) in which the similarity of the hydroxylation pattern of the benzene ring to that of the flavonoids should be noted); the xanthones (euxanthone, 53, Billet, 1966); and the stilbenes (e.g. resveratrole, 54) which are heartwood constituents. Mangiferin is a C-glucoside (5.3) of a xanthone (Billet *et al.*, 1965; Nott and Roberts, 1967).

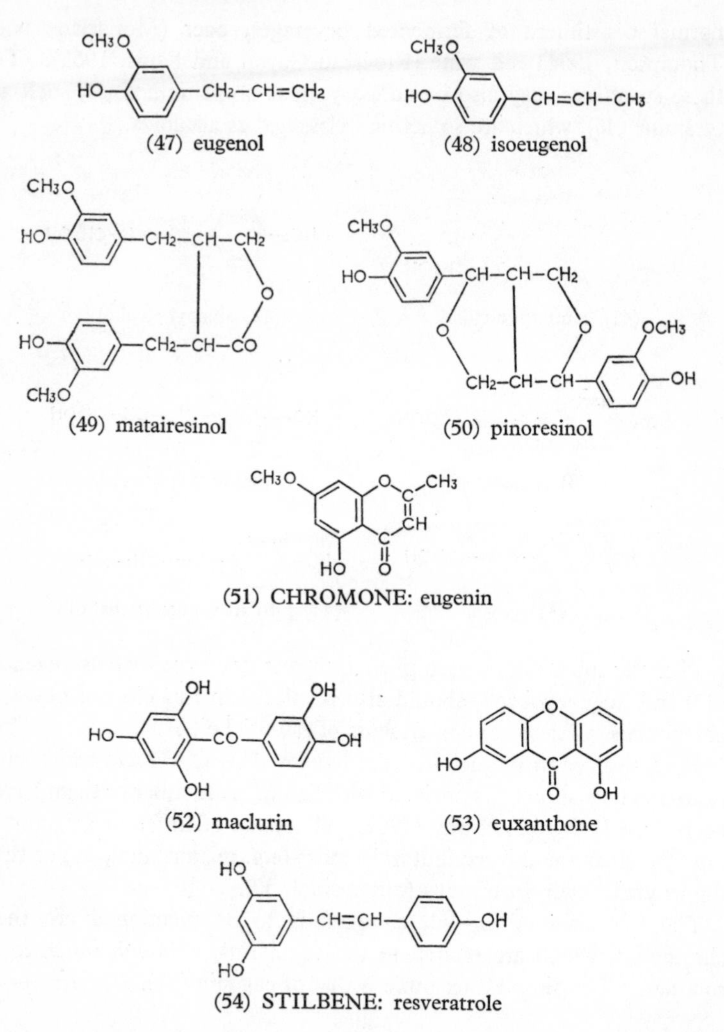

(47) eugenol

(48) isoeugenol

(49) matairesinol

(50) pinoresinol

(51) CHROMONE: eugenin

(52) maclurin

(53) euxanthone

(54) STILBENE: resveratrole

1.10. *Benzoquinones, naphthaquinones and anthraquinones*

Benzoquinones are common in the fungi, but in higher plants only a few, e.g. 2,6-dimethoxybenzoquinone (55), have been reported. However, ubiquinone (56), very well known in the animal kingdom as well as in the plant kingdom, can be regarded as belonging to this group. Its function as an electron-transport agent in oxidoreduction systems is well known. Naphthaquinones (e.g. juglone, 57) are relatively rare.

It is the anthraquinones which are most numerous and most widely distributed both in higher plants and fungi. Emodin (58) is one example of more than forty known members of this group.

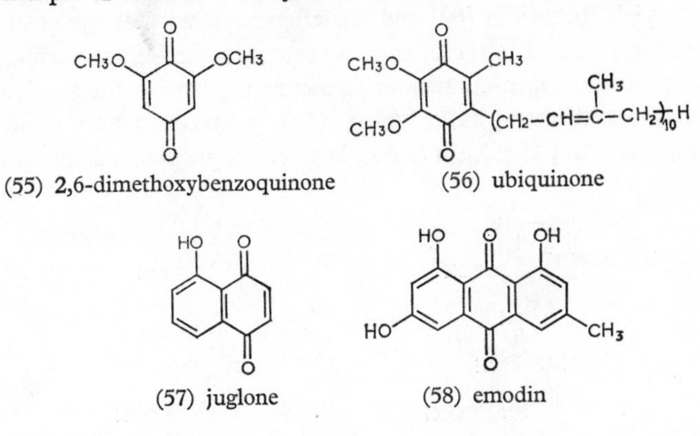

(55) 2,6-dimethoxybenzoquinone

(56) ubiquinone

(57) juglone

(58) emodin

1.11. *The biflavonyls*

These are C_{30} dimers either of apigenin (Table 1) or of a methylated derivative of this flavone. Except very rarely, these are confined to the gymnosperms. Among the dozen or more known members of this group, ginkgetin (59) is the most familiar.

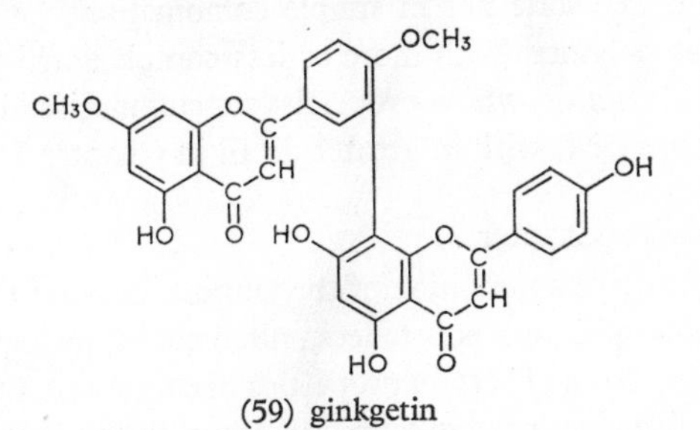

(59) ginkgetin

1.12. *The betacyanins and betaxanthins* (C_{18})

The betacyanins are red pigments responsible most notably for the colour of the red beet (*Beta vulgaris*). They are uniquely present in the Centrospermae.* Owing to their resemblance to the anthocyanins (e.g. their absorption spectra) and the presence in them of two atoms of

* But also in the Cactaceae, which are now recognised as being included along with the Centrospermae within a wider order. (Translator's note—E. C. B-S.)

nitrogen, they were for a long time known as 'nitrogenous anthocyanins' (Glennie and Sauvain, 1952). Their structures have recently been elucidated (Mabry *et al.*, 1963) as glycosides which, when hydrolysed with acid, yield betanidin (60) and are often acylated (Minale *et al.*, 1966). Closely related to these, and also found in the Centrospermae, are some yellow nitrogen-containing pigments, the betaxanthins. The structure of one of them, indicaxanthin (61), has recently been established (Piatelli *et al.*, 1965). It is not, however, a phenolic compound.

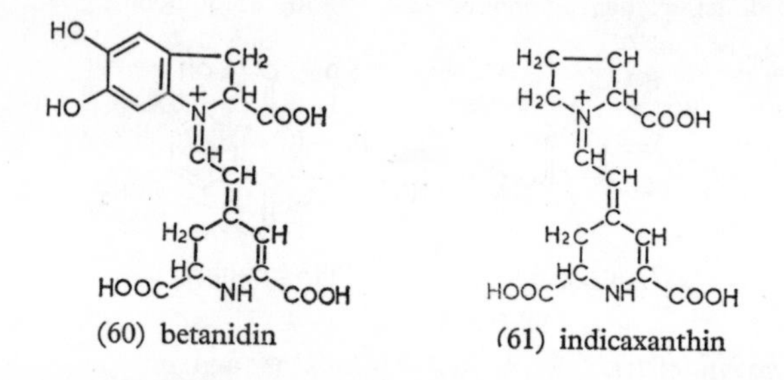

(60) betanidin (61) indicaxanthin

Phenolic compounds present in polymeric form

1.13. *General considerations*

There are two types of phenolic compounds present in plant tissues neither in the free state nor in simple combinations (as glycosides or esters), but as polymers with more or less complex structures. These are lignins and tannins, whose properties are summarised in this section. The tannins are dealt with in greater detail in Chapter 7.

1.14 *The tannins* (Chapter 7)

It is not easy to give a definition of the tannins, because this term comprehends an assemblage of substances united not by analogy of structure, but by the possession of certain properties in common. Etymologically, tannins are substances used in tanning leather, and they have, therefore, the property of converting raw hide into non-putrefying and not easily permeable leather. However, as Swain and Bate-Smith have pointed out (1962), the term tannins is broadly used in plant chemistry to denote a large number of substances widespread in plants, having properties similar to those of commercial tanning agents, but whose ability to tan hides has never been put to the test. These authors define tannins as phenolic compounds having molecular weights between 500 and 3,000, and which, as well as having the classical properties of phenols, precipitate alkaloids and gelatin and other proteins.

Depending on their molecular structure, tannins are customarily divided into hydrolysable and condensed tannins. Hydrolysable tannins are complex molecules containing ester type linkages which yield, on hydrolysis, a sugar residue and a phenolic residue consisting either of gallic acid (62) or ellagic acid (63), which is a dimer of the former. Bate-Smith (1956) has described a method of identifying ellagic acid by paper chromatography and has shown that this acid is frequently present in acid hydrolysates of leaves.

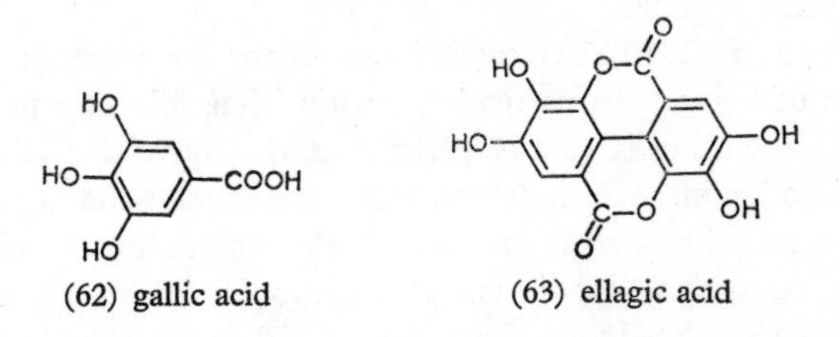

(62) gallic acid (63) ellagic acid

The study of the condensed tannins is actually less advanced, although they are, in fact, more important. It appears to be well established that they are formed by the polymerisation of molecular units having the general structure of flavonoids, the most important of which are flavan-3-ols (catechins) and flavan-3,4-diols (leucoanthocyanidins, Table 2). The copolymerisation of catechins and leucoanthocyanidins is also possible (Forsyth and Roberts, 1960), leading to biflavans. These are at the present time the object of intensive research (7.7).

The properties of the condensed tannins depend upon the nature of the molecular units of which they are composed, the mode of linkage of these units, and above all on the molecular weight of the condensation product. As an approximation, the tanning properties of a given specimen, that is its aptitude to combine with proteins, increases from the dimer to the decamer, and thereafter falls away. When the molecule gets too large it loses the power to combine firmly with protein; it may in fact become insoluble.

1.15. *The lignins*

Lignins may be defined as the non-carbohydrate moiety of all membranes. Their composition varies according to the kind of plant from which they are extracted and the conditions of extraction (Pearl, 1967). The chemical structure of lignin is not, however, well enough known to be able to characterise the differences in terms of the origin of the plant material.

Recent work (Barnoud, 1965) suggests that lignin results from the polymerisation of a C_6—C_3 compound corresponding essentially to coniferyl alcohol (41) and its methoxylated derivative sinapyl alcohol (42). It is now regarded as unlikely that, like polysaccharides (e.g. cellulose), the lignin polymer possesses an ordered structure. For instance, it is not possible to hydrolyse lignin, as one may hydrolyse a polysaccharide and obtain a simple, well-defined product. It has to be recognised that several different kinds of unit are involved in the constitution of this polymer.

Freudenberg (1959) has carried out important research into the structure of lignin. In particular, he has been able to reproduce *in vitro*, in presence of laccase or peroxidase, the conversion of coniferyl alcohol into lignin. He also showed that the condensation of coniferyl alcohol occurs by several oxidative pathways leading to dimers of different structures, for instance dehydrodiconiferyl alcohol (64), pinoresinol (50, already mentioned) and the β-coniferyl ether of guaiacylglycerol (65), lignin being formed by the eventual condensation of these dimers, which represent the true building stones of the polymer. This theory accounts for the complexity of the lignin structure although it is based on a severely restricted number of basic units.

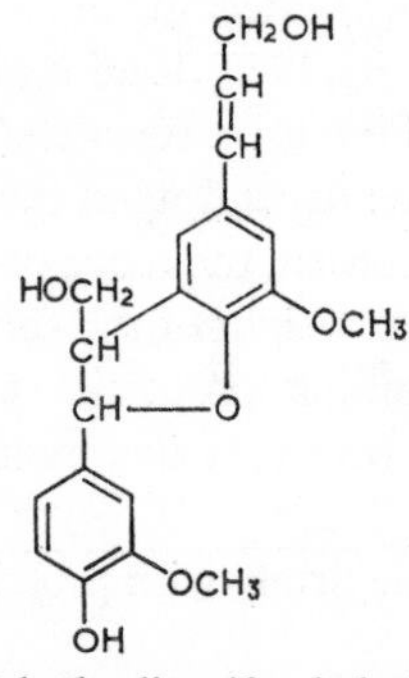

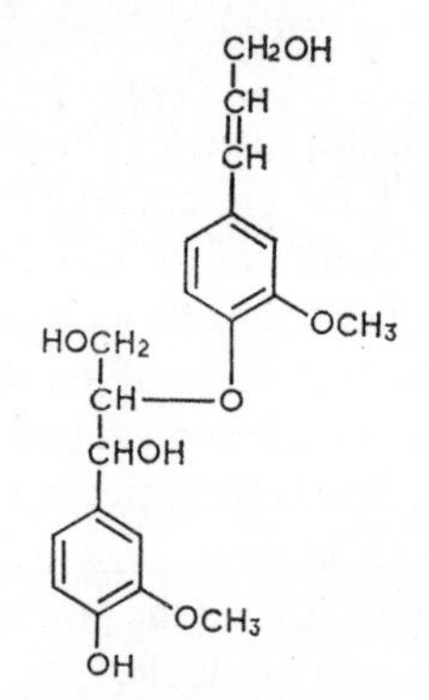

(64) dehydrodiconiferyl alcohol

(65) guaiacylglycerol β-coniferyl ether

Terminology and nomenclature of the phenolic constituents

1.16. *Terminology*

The word 'flavone' is derived from the Latin *flavus*, yellow. The parent substance of the flavonoid class, flavone (66), does occur naturally but is not itself phenolic.* The term 'flavonoid' was introduced by Geissman and Hinreiner (1952) to designate all the pigments having a C_6—C_3—C_6 skeleton analogous to that of the flavones (also including the anthocyanins). Sometimes the cinnamic acids and the coumarins are included within the flavonoids.

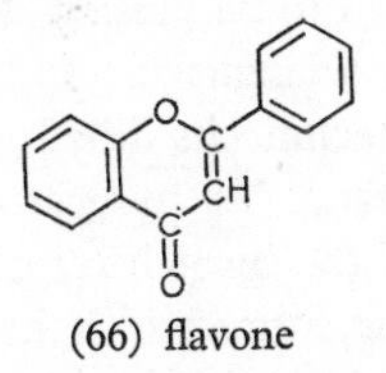

(66) flavone

Sometimes a distinction is made between anthocyanins, red and blue pigments, and anthoxanthins, which are yellow or 'white' petal pigments. The latter name derives from the Greek *anthos*, flower and *xanthos*, yellow.

'Chalcone' comes from the Greek *chalcos*, copper and 'aurone' from the Latin *aurum*, gold.

The name 'anthocyanins' is certainly the oldest, because it was introduced by Marquart in 1835. It comes from the Greek *anthos*, flower and *kyanos*, blue. It is used very broadly in the expression 'anthocyanin pigments' without regard to their nature, whether glycosidic or aglycone, i.e. whether they are anthocyanins or anthocyanidins.

The name 'leucoanthocyanin' lends itself especially to confusion. It is used to denote the totality of substances capable of conversion into anthocyanins (*sensu lato*) when heated with mineral acid, whether they are monomeric (i.e. leucoanthocyanidins) or polymeric, (i.e. tannins, 7.6).

1.17. *Principles of nomenclature*†

The terminology of secondary plant constituents has grown up somewhat independently of the strict rules which have now come to be accepted in academic chemistry. Following such precedents as quercetin and rutin, English and German usage has tended to adopt the termination *-in* when naming substances at the time of their first isolation, when nothing may be known about their structure other than that

* For many years, and even occasionally nowadays, the term 'flavone' has been used in France to refer to plant phenolic constituents generally. (Translator's note—E. C. B-S.)

† This section, for which the translator is responsible, replaces the original which deals for the most part with difficulties of usage in the French language.

they do not contain nitrogen. Departure from this practice has led to anomalies such as kaempferol and umbelliferone, the one a relative of quercetin, the other a relative of aesculetin, a coumarin.

French usage, on the other hand, has tended to employ the termination *-ol* for such substances when, as in the case of phenolic constituents, hydroxyl groups are known to be present, and this has sometimes led to confusion in the English language. In chemical abstracts catechol (Fr.) often appears for catechin, the dihydroxybenzene catechol (Eng.) being known still in France as pyrocatechol. French usage also recommended the use of the term heteroside for glycoside, and the termination-*oside* for those forms which in English are distinguished from aglycones by the insertion of *-id-* into the name of the aglycone, e.g. cyanin, the glycoside, cyanidin the aglycone. In the French original of this book, a plea is made for the regulation of the French nomenclature of the phenolic constituents, either by the adoption of the conventions now accepted in other languages, or by a rigorous application of the more internationally correct rules proposed by such modern French authors as Sannié and Sauvain (1952).

2

Chemistry of Phenols. Application to Natural Products

Introduction

It is considered necessary to devote a chapter in the early part of this book to a description of the chemical properties of phenols interpreted in terms of their electronic structure, especially as they determine the behaviour of the phenols in natural products. Such a treatment is useful from two points of view. First, an understanding of the chemical properties of phenols will be helpful in the interpretation of their biological properties; and, second, because the chemical properties are constantly being employed in natural product chemistry both in their detection on chromatograms and in their quantitative analysis.

Research workers interested in phenolic compounds are frequently specialists in biology or plant physiology, and do not always possess a deep knowledge of chemistry. For this reason, the present chapter, which deals with some of the simpler principles of chemistry, will appear rather elementary to some readers.

In drafting this chapter, the author has had much help from Arnaud's *Cours de Chimie Organique* (1964) and from the textbooks of Fieser and Fieser (1956), Cram and Hammond (1963) and Geissman (1965). For the relationship between structure and reactivity of the natural phenolic compounds, the article by Thomson (1964) has been consulted.

Questions of the synthesis of natural products are touched upon, because such questions are not dealt with in most of the recent work on phenolic compounds, and the research workers for whom this book is likely to be of interest will be concerned with this aspect of phenolic chemistry. However, the compounds used as reference substances for the purpose of identification are most often likely to be of natural origin.

It is well known that the functional group in phenols is a hydroxyl group attached to one of the carbon atoms of a benzene ring (aromatic radical). The formula of the simplest member, ordinary phenol, is C_6H_5OH, represented by the structure (1), or more usually (2). These structures are therefore analogous to those of the alcohols, in which an OH group is attached to an aliphatic radical. In the case of other

(1) (2)

functional groups of organic chemistry, the properties are not always the same in the aliphatic and aromatic series (e.g. the amines and halogen derivatives) but they are not sufficiently different for there to be any distinction made between the members of the two series. A distinction is made in the case of the 'alcohols' and 'phenols' because the phenolic hydroxyl group is profoundly affected by the presence of the benzene ring. When studying the chemical properties of the phenols, it is necessary to consider both the effect of the benzene group on the properties of the phenolic hydroxyl and the effect of the latter on the properties of the benzene ring to which it is attached.

Relationships between electronic structure and chemical properties of phenols

2.1 *The covalent bond and the electronic structure of benzene*

In order to be able to understand the effect of the benzene ring on the properties of the OH group, it is necessary first to consider some fundamental aspects of the electronic structure of the benzene ring itself.

The type of bond most concerned in the reactivity of the benzene ring is the covalent bond, in which two electrons are shared between the atoms involved. In the simplest case, the electron doublet common to the two atoms is strictly confined to the immediate neighbourhood of the bond axis. Such electrons are known as σ electrons, implying a σ bond of low reactivity and are therefore stable. Multiple bonds, especially double bonds $\left(>C=C< \right)$, are different, in that they are constituted by a σ bond of the same kind as that just described and a π bond formed by the sharing of two electrons which are much less strictly confined to the neighbourhood of the bond axis, because they are further from the centre of the nucleus. The π electrons can be displaced within a molecule from one bond to another, an event which is especially prone to occur when conjugation is present, that is an alternation of single and double bonds, as in the benzene ring. The double bond, involving 'mobile' or 'disposable' electrons, is much more reactive than the single bond. It constitutes a point of unsaturation at which addition reactions can take place, and also a point of diminished resistance in a carbon chain at which rupture can occur.

The four bonds of a carbon atom singly bonded do not all lie in one plane; they form an angle of 109° 28′, since they are associated in space towards the apices of a regular tetrahedron in which the carbon atom occupies the centre. From this it follows that a chain of carbon atoms joined by single bonds does not lie in a plane. On the other hand, in the case of a double-bonded carbon atom, the double and two single bonds lie in the same plane. From this it follows that a molecule consisting of conjugated double bonds is planar. The benzene ring is such a molecule.

In the case of the natural flavonoid compounds, consisting of two benzene rings united by an element containing three carbon atoms, the molecule will be planar or non-planar depending upon the saturation or unsaturation of the 3-carbon chain. The flavonols (3), for example, have a planar structure because of the double bond between C_2 and C_3; but the flavanonols (4) are non-planar, the two benzene rings not being

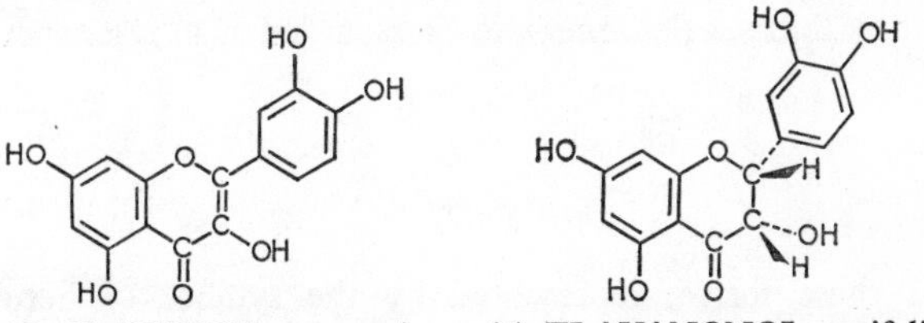

(3) FLAVONOL: quercetin (4) FLAVANONOL: taxifolin

in the same plane. In the structure for taxifolin (4), the bonds represented by a dotted line lie behind, those represented by a thick line in front of, the plane of the paper.

Having now dealt with some of the general properties of the chemical bond, the next step will be to consider the particular structure of the benzene ring.

When the structure of benzene is written according to the formula of Kekulé (5), comprising three double and three single bonds, the actual properties of the molecule are not truly represented. In actual fact, certain experimental facts (physical and chemical properties) indicate that the six bonds are identical, possessing characteristics intermediate in some respects between those of the single and those of the double bond. As one example, the length of a single C—C bond is 1·54 Å, that of a double bond 1·34 Å, but each of the six bonds of benzene has the same length, viz 1·39 Å. This peculiar structure of benzene is due to the presence of 6 π electrons which are not localised in pairs to produce typical double bonds, but are 'delocalised' over the whole of the ring, creating an identical electron density for each bond.

The conventional system for representing a covalent bond by a single line provides for the representation of single, double and triple bonds, but not for situations intermediate between them, as is the case with benzene. There is no really satisfactory graphical solution for this problem. One possibility is to depict the double bonds with dotted lines, as in (6).

(5) (6)

The most correct representation of benzene would be one which took into account all the different possibilities of localisation, two by two, of the π electrons (7). There is a state of so-called mesomery between all the limiting forms. Benzene is to be regarded as a 'resonance hybrid'

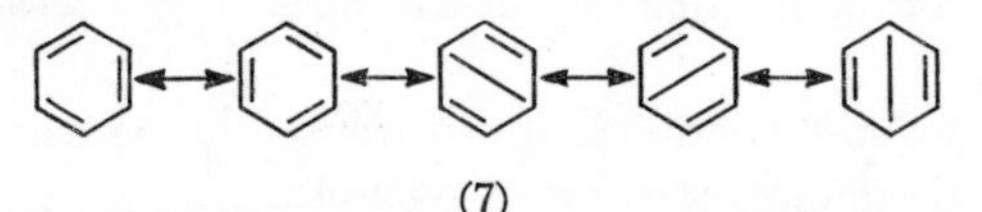

(7)

between these forms, represented by the symbol ↔ between the different formulae. But it would be incorrect to regard benzene as a mixture of different molecules corresponding with the several different formulae, or that it exists at different times in any of the different states. Benzene has one structure, but it is not possible to provide a single graphical representation of it which covers all its different properties.

One important consequence of mesomery is that it imparts an increased stability to the molecule. For one thing, additive reactions are much more difficult than is the case with ethylenic hydrocarbons. The benzene ring often behaves like a saturated system, undergoing substitution reactions (replacement of hydrogen by another chemical group) of great importance. Finally, the benzene ring is seldom ruptured in the course of chemical or biological reactions.

2.2 *The electronic structure and the acid character of the phenolic function*

In phenol, which contains a hydroxyl group attached to a benzene ring (8), interaction takes place between the mobile electrons of the ring and the unshared electrons of the doublets of the oxygen atom. This interaction allows the possibility of displacement of these electrons, in other words a mesomeric effect which is expressed by writing different limiting formulae corresponding to the different states of localisation:

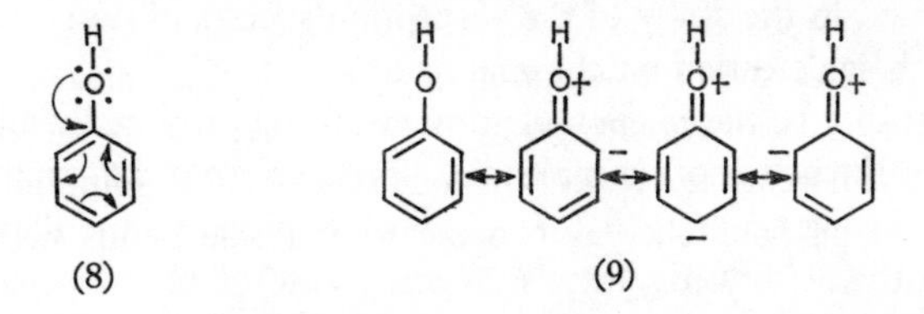

(8) (9)

This mesomeric effect results in a decrease in the electron density of the oxygen (appearance of a positive charge). Normally, oxygen possesses two pairs of free electrons and two unpaired electrons providing for two covalent bonds. If, because of the structure of the molecule, an oxygen atom loses an electron belonging to an unshared doublet, the atom acquires a positive charge, but it will then have three unpaired electrons and consequently has at its disposal three covalent bonds. The implications of a negative charge appearing at the apex of the ring, shown as one of the limiting formulae of phenol (9), will be discussed later (2.7).

The impoverishment of the electron density of the oxygen atom results in the attraction towards the oxygen atom of the electrons forming the bond between hydrogen and oxygen in the phenolic hydroxyl group. Consequently the rupture of the OH bond (10) is facilitated by the presence of the benzene ring, whilst the rupture between C and O (11) is made more difficult.

The ease of rupture between O and H makes the hydrogen atom of

(10) (11)

the phenolic function labile, conferring upon it acid properties, although those of a weak acid. Like all acids, the phenols are partially dissociated in water, at any rate to the extent that they are soluble. In the case of phenol itself, the dissociation constant is of the order of 10^{-10} ($p_K = 10$), compared with 10^{-4} to 10^{-5} for carboxylic acids and only 10^{-16}-10^{-19} for alcohols. The mobility of the hydrogen atom of the OH group constitutes the essential difference between alcohols and phenols; it is this which determines their special properties.

The phenols are therefore neutralised in alkaline solutions; they assume the form of 'phenolates', strongly dissociated like all salts, with the formation of ions of the type $C_6H_5O^-$ and with spectral properties different from those of the corresponding phenol. This property is widely used in the study of the absorption spectra of natural products and for their detection on chromatograms.

The acidity of the phenolic hydroxyl can be profoundly affected by the general structure of the molecule. For example, 2,4,6-trinitrophenol (12) is a strong acid ($p_K = 0{\cdot}71$), known as picric acid. These differences in the acidity of phenols are made use of in the separation of natural phenolic compounds with the help of alkaline solutions (3.2).

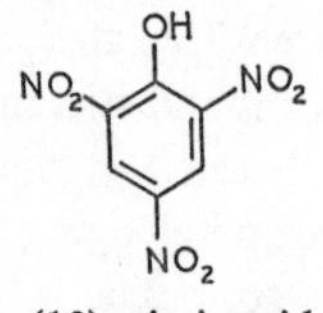

(12) picric acid

For example, in the case of the flavonoids, the presence of a CO group (flavone, flavonol, etc.) which can 'accept' electrons from the benzene ring increases the polarisation of the OH group, hence the mobility of the hydrogen atom, and hence its acidic properties. (The NO_2 groups of picric acid have a similar effect.) In a mixture of phenolic compounds, a weakly alkaline solution (sodium carbonate or bicarbonate) will only neutralise the most acidic of the phenols, converting them into salts soluble in water. The free phenols, being less acidic, will be selectively extracted by organic solvents (e.g. ether, ethyl acetate).

Properties of the phenolic hydroxyl group

2.3. *Formation of hydrogen bonds*

Like the alcohols, phenols are the site of intermolecular association by way of hydrogen bonds. In complex phenolic molecules, such bonds can also be formed intramolecularly. The effect of these bonds is to modify numerous physical properties, such as melting and boiling points, solubility, and the ultraviolet and infrared spectra.

The source of the hydrogen bond is essentially electrostatic. It occurs when the hydrogen atom is combined with an electronegative atom such as oxygen, causing polarisation of the O—H bond. The partial denuding of the hydrogen bond of its electron results in its seeking to procure one from an oxygen atom of another molecule having a high electron density. This bond is represented by a dotted line (13).

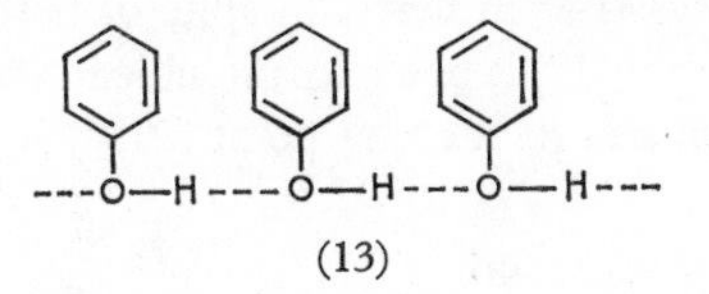

(13)

The bond so formed is much weaker than a covalent bond, and molecules so associated recover their autonomy when they enter the gaseous phase. The hydrogen bond, however, raises the boiling point because it increases the energy necessary to free a molecule from interaction with its neighbours.

In the case of natural phenols (Thomson, 1964), intramolecular hydrogen bonds are common, notably when an *o*-hydroxyacetophenone (14) grouping is present. This is especially evident in the case of the flavone derivatives hydroxylated in position 5 (e.g. quercetin (3) and taxifolin (4)). Similarly, infrared spectroscopy has revealed the presence of hydrogen bonds between the OH group in position 3 and the heterocyclic oxygen in the case of the catechins (15) (7.7).

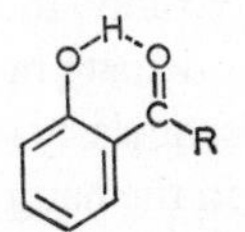

(14) *o*-hydroxyacetophenone

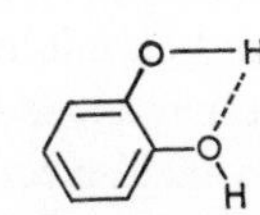

(16) catechol

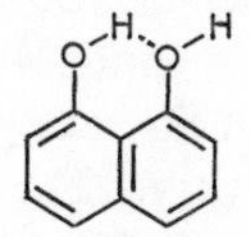

(17) 1,8-dihydroxynaphthalene

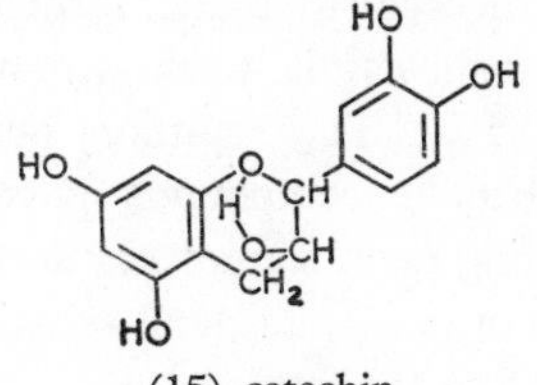

(15) catechin

Hydrogen bonding generally reduces the reactivity of the phenolic groups, for example their solubility in alkali and their aptitude to form esters and ethers. On the other hand, hydrogen bonds are stronger when they are concerned in the formation of a 6-membered 'ring' (e.g. in the case of *o*-hydroxyacetophenone (14) or 1,8-dihydroxynaphthalene (17)) than when the 'ring' so formed has only 5 members (e.g. in catechol (16)).

Intermolecular hydrogen bonds affect certain physical properties of phenolic compounds. More particularly, they raise the melting point and lower solubility. For instance, in spite of having three hydroxyl groups, phloroglucinol (18) is much less soluble in water than resorcinol (19) with only two, and its melting point (219°) is much higher than

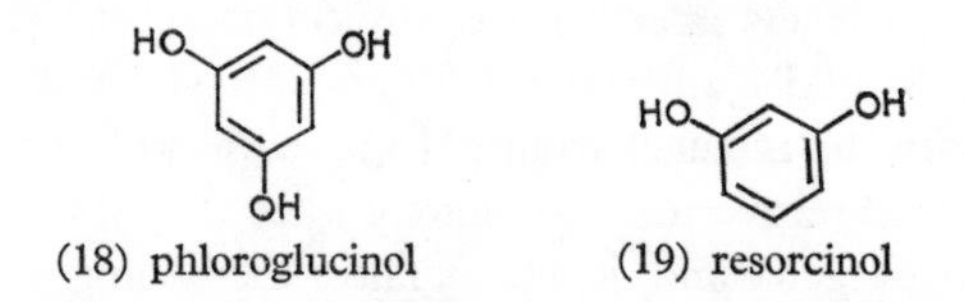

(18) phloroglucinol (19) resorcinol

that of resorcinol (118°). Furthermore, these bonds may make it more difficult to purify phenols. They tend to form hexagonal structures arising out of six phenolic groups linked by hydrogen bonds forming cavities (clathrates) giving inclusion complexes with a large number of organic molecules, especially with those of the solvent used in the purification process.

2.4. *Formation of complexes with metals*

Considerable use is made of the ability of phenolic compounds to form complexes with metals (iron and aluminium in particular) in revealing spots on chromatograms, in estimation, and most particularly in spectrophotometry. These complexes are also concerned in the natural colours of plants (8.10).

Although there are very many references in the literature to the use of these complexes, the mechanism of their formation has been relatively little studied. The most substantial work, without any doubt, is that of Jurd and Geissman (1956). These authors pointed out that a large number of natural phenolic compounds possess certain structural groups capable of forming metal complexes: *o*-dihydroxyphenols (20), 3- and 5-hydroxychromones (21, 22), and *o*-hydroxycarbonyl residues (23) that can be attached to the former.

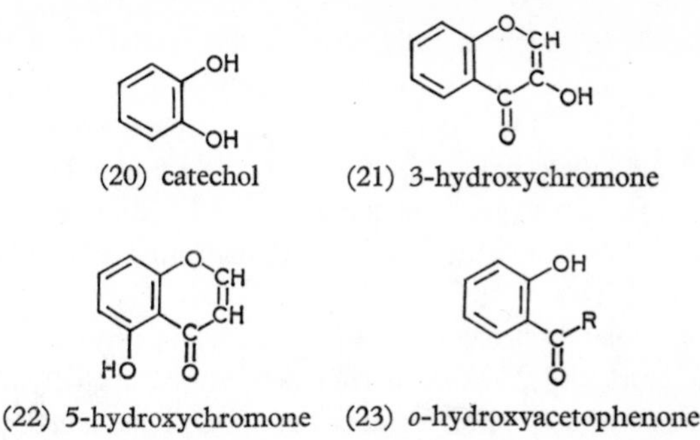

(20) catechol (21) 3-hydroxychromone

(22) 5-hydroxychromone (23) *o*-hydroxyacetophenone

Jurd and Geissman also showed that the configuration of the molecule as a whole was concerned in complex-formation. For instance, aluminium chloride has much less effect on the absorption spectrum of catechol (20) than on that of 3,4-dihydroxychalcone (24), which has a CO group conjugated with the catechol group.

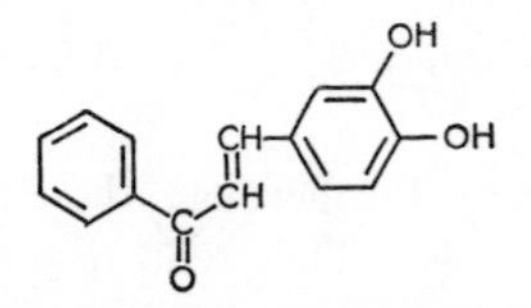

(24) 3,4-dihydroxychalcone

The modification of the spectrum is also considerable in the presence of sodium acetate, i.e. under mildly alkaline conditions. This indicates the existence of a relationship between the formation of the complex and the ionisation of at least one of the phenolic functions. In the case of 3′,4′-dihydroxyflavone (25), Jurd and Geissman interpret the formation of the complexes (28) or (29) on the basis of the ions (26) or (27) formed under alkaline conditions.

The complexes capable of being formed by the different structures (20) to (23) can be written as represented in (30) to (33) (Geissman, 1956). Fieser and Fieser (1956) write the formula for the complex formed by aluminium chloride and an *o*-hydroxycarbonyl function as in (34). Markham and Mabry (1968) have shown that the complex with aluminium chloride of type (30) is more stable in the presence of HCl that those of type (31) and (32). They use this property for the detection of *ortho*-dihydroxy groups (3.16 and 5.12).

In the case of natural complexes, Bayer (1966) attributes the blue colour of the cornflower (8.10), or 'protocyanin', to a complex of two

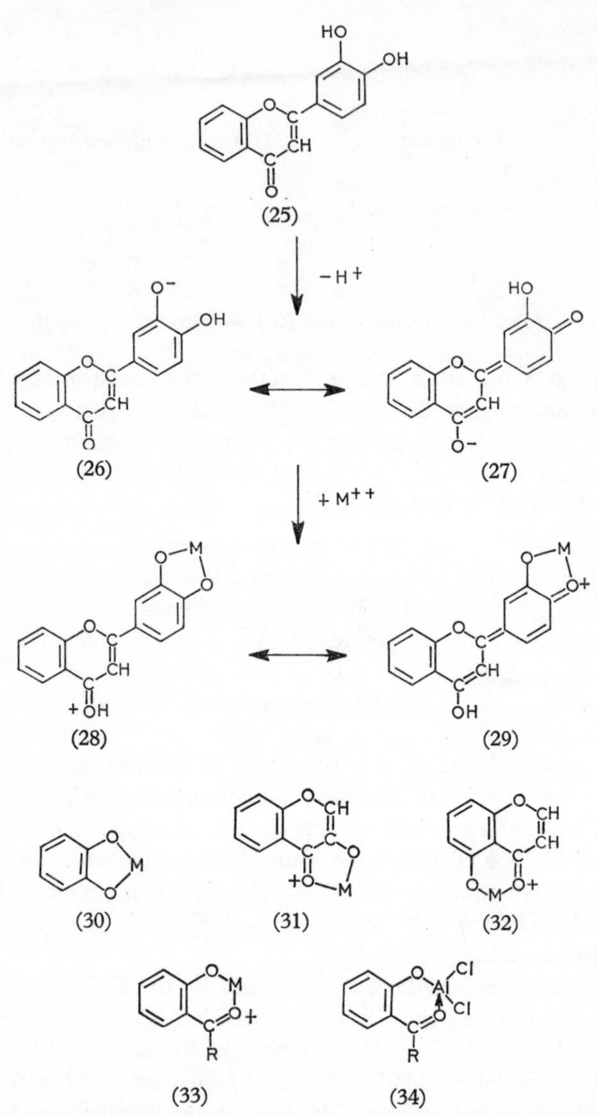

molecules of cyanidin linked through their *o*-diphenol groups with a single Al^{3+} or Fe^{3+} ion (34a). Harper (personal communication) has shown that the structure of metallic complexes of the anthocyanins—to be more precise, the number of anthocyanin ions per metal ion—varies with pH.

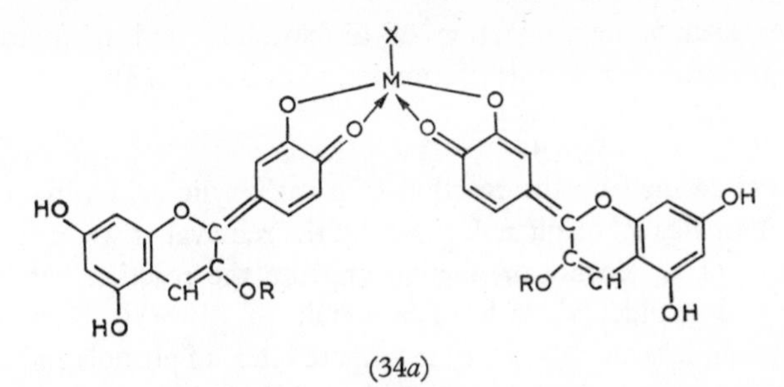

(34*a*)

As in the case of hydrogen bonds, the stability of the complexes depends upon the structure of the groups concerned. The complexes formed with *o*-dihydroxycarbonyl groups are more stable than those formed with *o*-diphenols. For this reason, iron and aluminium complex better with flavonols than with dihydroxylic anthocyanins. A fact well known to horticulturalists can be interpreted on this basis: when hydrangeas are watered with iron or aluminium alum, blue flowers are produced, but in the case of roses the result is negative, because in this case the metallic ions are perpetually combined with flavonols, and also with citric acid (Bayer, 1966; 8.10). In the laboratory, it is customary to use aluminium in the form of chloride ($AlCl_3$) for the formation of complexes; also lead, in the form of neutral lead acetate. The latter forms precipitates, the colours of which vary with the nature of the phenolic compound (Geissman, 1955). This property is used both in the removal and in the purification of these substances (3.3).

It is in connection with absorption spectra that metallic complexes are most often made use of, since they produce changes which are quite characteristic of the substances studied. Use is also made of them in the development of chromatograms; it is actually possible to

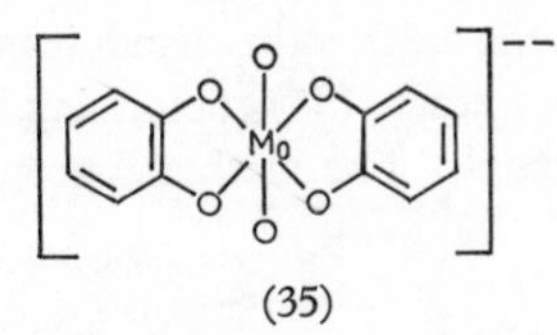

(35)

impregnate the paper with metallic salts and thereby obtain characteristic shifts in R_f. They are also made use of in the estimation of phenolic compounds. For instance, molybdenum in the form of sodium molybdate, forms a complex (35) with dihydroxyl and trihydroxyl groups with λ_{max} 300-320 nm, allowing photometric determination (Swain and Goldstein, 1964). Finally, a method has been described for the separate determination of di-(catechol) and tri-(pyrogallol) hydric groups using complex formation with iron (2.13).

2.5. *Formation of esters and ethers*

Esters are formed by the reaction of a carboxylic acid with the OH group of an alcohol or phenol, either by the removal of water from two molecules of an alcohol or phenol, or from the reaction between an alcohol and an aldehyde to form an acetal.

It is possible in the laboratory to prepare esters of phenols and organic acids, but not from the acids themselves; it is necessary to use an acid chloride or anhydride:

$$C_6H_5-OH + CH_3-C(=O)-Cl \longrightarrow CH_3-C(=O)-O-C_6H_5 + HCl$$

Unlike alcohols, it is not possible to esterify phenols with mineral acids.

In nature, esters of phenols are practically unknown, apart from certain compounds of gallic acid (1.14, structure 62) such as ellagic acid (1.14, structure 63) which contains two ester bonds. In the case of acylated anthocyanins which contain organic acids (usually cinnamic acids), Harborne (1964) showed that the acyl residue is esterified with a hydroxyl group of a glycosidic sugar residue and not with a phenolic hydroxyl (6.2). Coumarins might equally well be regarded as internal esters of *o*-coumaric acids. Generally speaking, the esters which come under consideration in the study of natural phenolics are those arising from the combination of an acid phenolic group and an alcoholic hydroxyl of another molecule. The phenolic compound is not in these circumstances functioning as a phenol. Ester linkages are broken by saponification, that is by alkaline hydrolysis, usually in the cold.

The formation of ethers, on the other hand, is much more frequent in nature, especially that of ethers (36) comprising both a phenolic component and methyl alcohol. The methoxyl group (—OCH_3) is very

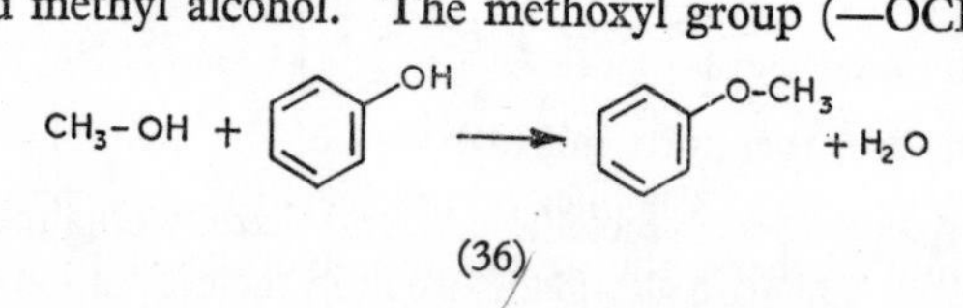

common, especially in the anthocyanins and phenolic acids. It is exceptionally stable, both chemically and biologically. In the laboratory, the OH group of the natural products can be methylated with methyl sulphate in acetone solution in the presence of sodium carbonate. This reaction is especially useful in determining the position of glycosyl residues in flavones and flavonols (5.10).

The bonds joining the aglycones and sugar residues in glycosides, which are called glycosyl bonds, are of the same type as those in ethers —C—O—C—, but they differ in their properties. They are, in fact, ruptured by heating with acid, whereas methoxyl groups are not affected by this treatment. The glycosyl bond is functionally an internal semi-acetal of the sugar molecule. A semi-acetal group (39) is formed by the interaction of an aldehyde (37) with an alcohol (38). In the case of the sugars (40 and 41), these two functions belong to the same molecule. In the formation of a glycoside, the semi-acetal (42) is transformed into an acetal (43) by the intervention of a phenolic or alcoholic hydroxyl group.

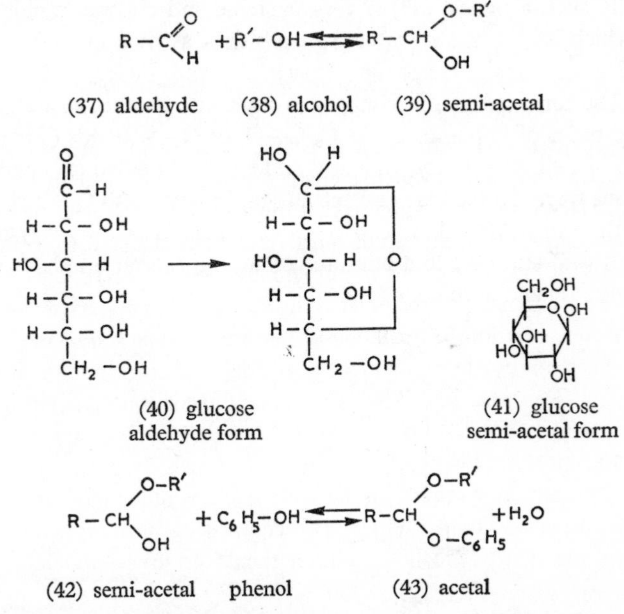

The acetals are, therefore, mixed ethers of the unstable hydrates of aldehydes (or ketones). Chemically, the acetalisation reaction is reversible, and unlike ethers, the acetals are hydrolysed by hot acids, a reaction much made use of in the study of natural glycosides.

2.6. *Oxidation*

As Thomson (1964) has pointed out, the oxidation of phenol has been rather neglected by organic chemists, probably because of its extraordinary complexity, despite the need for such information in the study of natural products. To take one example, the chemical or enzymic oxidation of the phenolic constituents of plant tissues results in browning, which is of great importance in food technology, especially in oenology. In another direction, advantage is taken of the ready oxidation of some phenols in employing them as antioxidants for the protection of oils and fats. Also in the biosynthesis of the phenolic constituents, whether these are monomeric like the flavonoids (8.5) or polymeric like lignin (1.15) and the tannins (7.8), oxidative mechanisms are involved.

The first step in the process of attack of oxidising agents on phenols is the removal of a hydrogen atom and the formation of a free radical having a singly-bound oxygen atom carrying an unpaired electron. Usually such a radical (45) is very unstable and changes rapidly into one which has, in the *ortho* or *para* position, a trivalent carbon atom carrying an unpaired electron (46) (Fieser and Fieser, 1956). The radical so formed can dimerise, or can react with another radical, forming, in order of importance, C—C, C—O, or O—O bonds (Thomson, 1964). In all probability, chemical and enzymic oxidations operate in the same way. In the case of catechol (44) (Forsyth and Quesnel, 1957; Forsyth *et al.*, 1960) these two types of oxidation lead by way of the radicals (45) and (46) to a mixture containing several tetrahydroxybiphenyls (47) and a quinone (48).

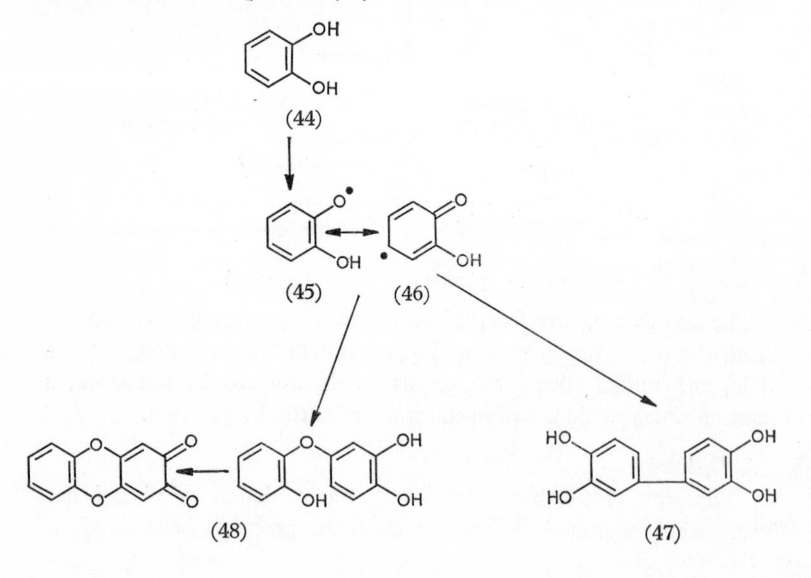

In the case of a monophenol, *p*-cresol (49), Thomson (1964) explains the oxidation, chemical or enzymic, by an analogous mechanism, either by way of a dimole with two phenolic functions (53) or a ketone (55). This mechanism corresponds to the oxidation of a monophenol by peroxidase in the presence of H_2O_2. In either case (diphenols or monophenols), further oxidation, or oxidation of complex molecules

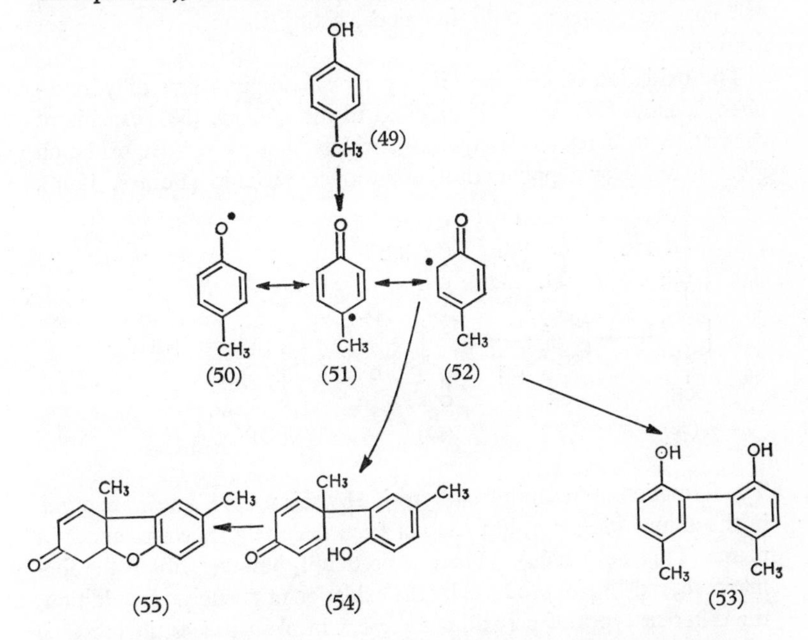

containing catechol residues, can lead to dark brown polymeric products. This kind of behaviour can account equally well for the browning of vegetable foodstuffs or for the formation in the plant of brown products such as lignin (1.15). Guyen *et al.* (1965) have shown that the autoxidation of quercetin in weakly alkaline conditions leads to a number of well-characterised polymers.

Enough has been said to support the claim made earlier that the oxidation of phenols is a subject of great interest about which all too little is known. Much more attention has been paid, however, to the enzymes involved in the oxidation processes (see Pridham, 1963). These enzymes are termed 'phenolases', 'phenoloxidases', 'polyphenoloxidases', or even 'O_2: oxidoreductases'.

It is usual to divide phenoloxidases into two classes: tyrosinases and laccases. There is debate as to mode of action of laccases. While the oxidation of *p*-diphenols is well understood, that of *o*-diphenols and even that of monophenols has yet to be elucidated (Burges, 1963; Conn, 1964). Tyrosinase type enzymes have two different kinds of activity: first, they convert monophenols into *o*-diphenols (cresolase activity); and second, they oxidise diphenols (catecholase activity). It has not been possible to isolate two different enzymes. It seems likely that the one enzyme possesses both functions, acting through two different centres.

The oxidation of tyrosine (56) by tyrosinase gives first dihydroxyphenylalanine (57), which is oxidised to the quinone (58), capable of cyclisation to a red derivative (59), which then by further oxidation gives a red-brown polymerisation product, melanin (Burges, 1963).

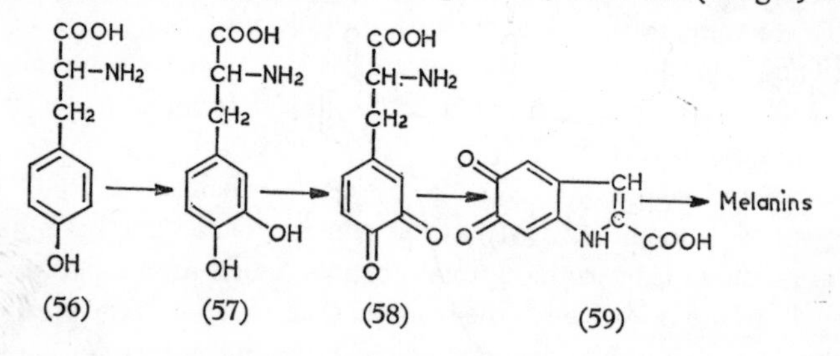

The formation of melanin from tyrosine by this route is known to occur in the animal kingdom, but has not been definitely shown to occur in plants (Thomson, 1964). There is no doubt, however, about the formation of red-brown products by the oxidation of phenols. In addition, the different types of peroxidases present in plant tissues, in presence of dissolved oxygen or hydrogen peroxide, oxidise phenols, especially the monophenols (49) to (55), just as they oxidise other organic substrates.

From the analytical point of view, oxidising agents are used for the determination of phenols and for their detection on chromatographic papers. The use of ammoniacal silver nitrate, of a mixture of phosphotungstic and phosphomolybdic acids (Folin-Denis reagent), and of alkaline permanganate in the cold for these purposes will now be described. It should also be mentioned that phenols reduce Fehling's solution.

Ammoniacal silver nitrate This reagent is the equivalent of a molecule of silver oxide, Ag_2O, which, in the presence of a reducing agent, is reduced to silver. It is widely used for the development of chromatograms, reducing substances being oxidised to give a dark brown spot.

The reaction is usually carried out by freshly mixing equal volumes of 0·1N $AgNO_3$ and 0·1N NH_4OH, at a temperature governed by the nature of the substances to be revealed (room temperature up to 100°). The time and temperature at which a substance reacts can give a clue to its identity. This reaction is very sensitive, but it lacks specificity, because so many different kinds of substances, especially sugars and organic acids, respond to it. In the case of natural phenolic compounds, more specific reagents are to be preferred (3.10).

Folin-Denis reagent This reagent consists of a mixture of phosphotungstic ($H_3PW_{12}O_{40}$) and phosphomolybdic ($H_3PMo_{12}O_{40}$) acids, which is reduced, concomitantly with the oxidation of the phenols, to a mixture of the blue oxides of tungsten (W_8O_{23}) and molybdenum (Mo_8O_{23}) (Swain, 1964). The coloration produced (λ_{max} between 725 and 750 nm) is proportional to the concentration of phenolic compounds. The reaction is very commonly used for the determination of total phenols in natural extracts.

Potassium permanganate Acid potassium permanganate (0·01 N) oxidises phenols in the cold, a reaction which provides the basis for a classical method of determining total phenols, much used especially in oenology. In the case of wine, the reaction is carried out in the presence of a large excess of indigocarmine, which serves in the first place as an indicator, since its colour changes from blue to yellow in the presence of an excess of permanganate; but it acts also as a 'limiter of oxidation', being less oxidisable than the phenols but more oxidisable than the other substances in wine susceptible to oxidation by permanganate (Ribéreau-Gayon and Peynaud, 1958). Phenolic compounds, at equivalent concentration, do not all reduce the same amount of permanganate. The result obtained represents an average value providing an index, the permanganate index.

Bate-Smith (1954) has reported that the oxidation of phenolic compounds by permanganate depends upon their nature: only *ortho*- and *para*-diphenols react, monophenols and *meta*-diphenols do not, whilst certain non-phenolic substances do react, e.g. cinnamic acid, by virtue of its double bond. For this reason, potassium permanganate is not regarded as a reliable reagent for the determination of total phenols. Over a range of plant tissues, there is no certain relationship between the permanganate index and the phenolic content. It is better to use the Folin-Denis reagent, which reacts with all phenolic functions, whether there are mono- or diphenols.

However, the permanganate index is still of great value in oenology for the determination of total phenols in red wine. In this instance, it is a question of comparing the composition of different samples of wine of closely similar origins.

Specific properties of the benzene ring of phenols

2.7. *Electrophilic substitution of the benzene ring*

In discussing its electronic structure, it has been made clear that benzene cannot be regarded simply as an ordinary unsaturated hydrocarbon. The exceptionally stable benzene ring undergoes substitution reactions (replacement of H by a substituent) more readily than it does addition reactions. The ease with which substitution can take place, without addition, is an essential characteristic of substances possessing a benzene ring, i.e. belonging to the class of aromatic organic compounds.

All these substitutions are the results of the attack upon the benzene ring of an electrophilic reagent, i.e. one deficient in electrons, constituting a positively charged group. This electrophilic reagent is formed from a molecule X-Y by the action of a catalyst Z. When it approaches the benzene ring (60), X^+ attracts the mobile π electrons and concentrates them on one of the apices of the ring. When the electron density on this apex is high enough, a bond can be formed between this carbon atom and X^+. The carbonium ion (61) formed as an intermediate is stabilised by loss of the proton H^+ and the final product of the reaction, (62), is formed. In the presence of H^+, YZ^- regenerates the catalyst Z.

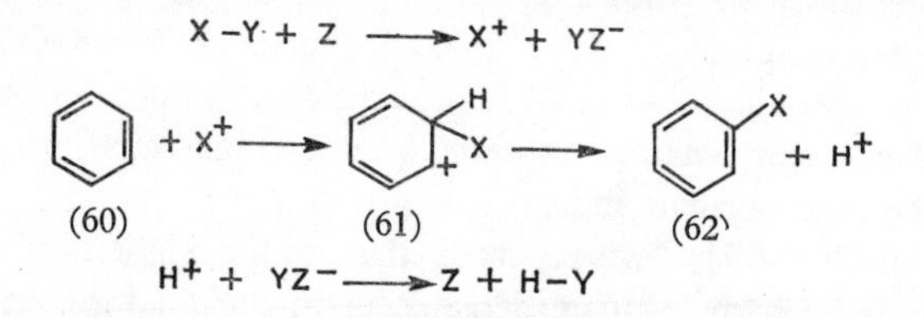

A carbonium ion (or carbocation) is formed by a carbon atom losing an electron and in consequence gaining a positive charge. Since this carbon atom has only three unpaired electrons, it can form only three covalent bonds. Conversely, a carbanion, negatively charged, is formed by a carbon atom possessing an extra electron, i.e. a doublet and three unpaired electrons; but like the carbocation, it can only form three covalent bonds.

The Friedel-Crafts reaction provides the classic example in organic chemistry of electrophilic substitution in the benzene ring. It consists in the substitution of a hydrogen atom of the ring by an aliphatic radical, brought about by the action of an aliphatic halide (e.g. methyl chloride, CH_3Cl) on a benzenoid compound (e.g. benzene itself, C_6H_6), in the presence of aluminium chloride, which plays the part of a catalyst and promotes the formation of the electrophilic radical CH_3^+.

Up to this point, the general principle of substitution has been illustrated by the particular instance of benzene. It is now important to consider the effect on the substitution reaction of one or more hydroxyl groups, and thereafter of that of other kinds of groups, already attached to the benzene ring.

In the case of phenol, the mesomeric situation has been depicted involving three limiting structures (63). These structures indicate that

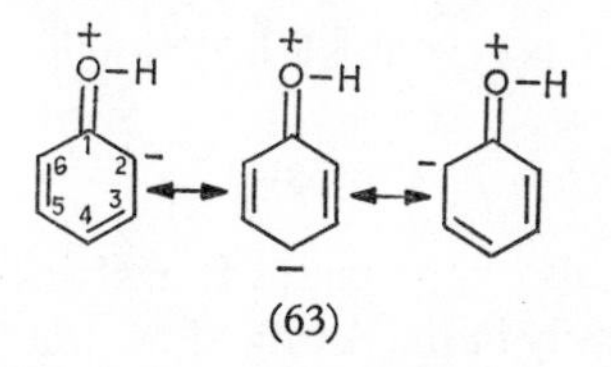

(63)

the electronic density is particularly high in positions 2, 4 and 6, i.e. *ortho* and *para* to the phenolic hydroxyl. It follows that substitution reactions, which require a strong electronic density, are favoured in these positions. In another sense, the hydroxyl group functions as a 'donor' of electrons; it increases the electron density on the benzene ring and enhances the ability of the ring, which is in effect 'active', to undergo substitution reactions. To sum up, substitution reactions are carried out more easily on phenol than on benzene, and they occur at positions 2, 4 and 6.

Coming now to polyphenols in which several OH groups are in alternate positions, substitution becomes very easy, especially so in the case of resorcinol (64) and phloroglucinol (65). Conversely, substitution is less easy in the case of catechol (66) and pyrogallol (67), where the OH groups are adjacent to each other. This is important because these last two residues are very common in nature, and especially so in the flavonoids.

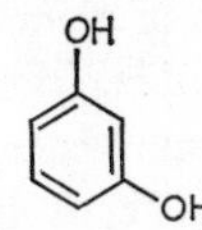

(64) resorcinol

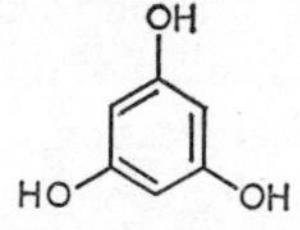

(65) phloroglucinol

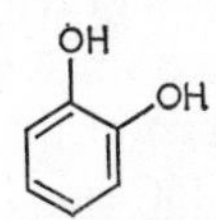

(66) catechol

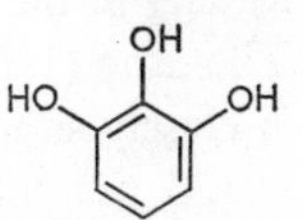

(67) pyrogallol

Other groups, e.g. the CO group, have an opposite effect. They behave as 'acceptors' of electrons. In the case of benzoic acid, the mesomeric form involves three limiting structures (68). In this case, the oxygen of the CO group 'accepts' a supplementary electron (whence the existence of a negative charge); it possesses therefore only one unpaired electron and can form only one bond. From these structures, it can be seen that the electron density, and therefore the reactivity, is decreased in positions 2, 4 and 6 while remaining normal in positions 3 and 5 (the *meta* positions). The end result is a 'deactivation' of the benzene ring and an orientation of substitution in the *meta* position.

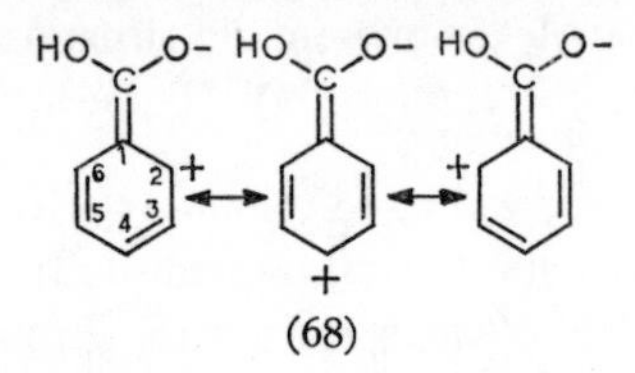

(68)

This last result is also important. It explains why substitution reactions occur more easily in the A ring of natural flavans (e.g. catechin, 69) than in that of flavones and flavonols (e.g. quercetin, 70). In each case the A ring has the phloroglucinol pattern of substitution, with one hydroxyl group engaged in the heterocyclic bond. In the case of catechin, the A ring is particularly easily substituted in positions 6 and 8, but in quercetin the A ring is 'deactivated' by the CO group. As an example, vanillin reacts (2.9) with catechin, but it does not react with quercetin. In either case, the B ring, with catechol type substitution, is less reactive; it does not react with vanillin.

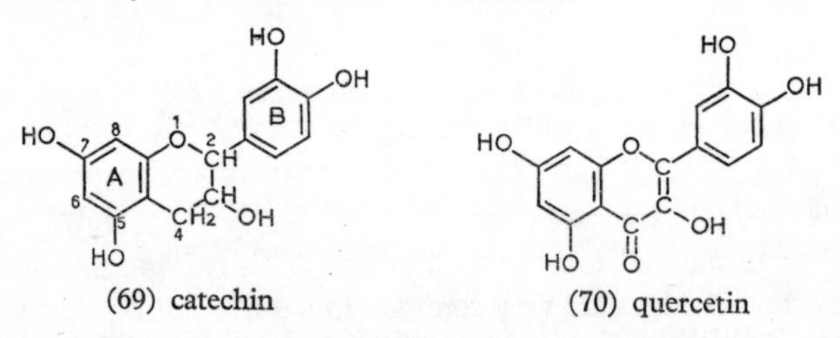

(69) catechin (70) quercetin

Substitution reactions undoubtedly occur *in vivo* in the biosynthesis leading to the formation of the various phenolic constituents of plants. Thomson (1964) remarks that the mechanism of the biological reaction is closely similar to that of the chemical reaction. An enzyme system would promote the formation of an electrophilic reagent effectively replacing the aluminium chloride which acts as a catalyst in the Friedel-

Crafts reaction. All the same, little is known of the ways in which electrophilic substitution reactions are involved in metabolic pathways. Such a reaction must, however, be involved in the polymerisation of flavans when they are converted into condensed tannins (7.8). In this polymerisation, carbon atoms 6 and 8 occupy a position of privilege, and in the case of the flavan-3,4-diols carbon atom 4 is exceptionally prone to form a carbonium ion.

In the section which follows a number of reactions are considered which are much used in the study of natural phenolic compounds, especially in their determination or in revealing them on chromatograms.

2.8 *Coupling with diazonium salt*

This reaction for phenols, leading to coloured derivatives, is much used for their detection on paper chromatograms. It is characteristic of primary amines (R—NH_2) which react with nitrous acid to give diazonium salts. In practice, nitrous acid itself is not employed, being unstable, but sodium nitrite, $NaNO_2$, which in the presence of hydrochloric acid reacts to form nitrous acid:

$$NaNO_2 + HCl \rightleftharpoons HNO_2 + NaCl$$

The diazotisation reaction may be written:

$$R{-}NH_2 + HNO_2 + HCl \longrightarrow R{-}N_2^+\,Cl^- + 2H_2O$$

In the aromatic series, the diazonium chloride, Ar—$N_2^+Cl^-$ is stable, reacting to give substitution reactions especially with the benzene ring of phenols as follows:

$$Ar{-}N_2^+\,Cl^- + C_6H_5{-}OH \longrightarrow Ar{-}N{=}N{-}C_6H_4{-}OH + HCl$$

This reaction, known as 'coupling', leads to the formation of 'azo dyes', the colours of which vary with the nature of the phenol.

In the laboratory, especially for the detection of phenolic acids, diazotised *p*-nitroaniline (71), diazotised sulphanilic acid (72) and

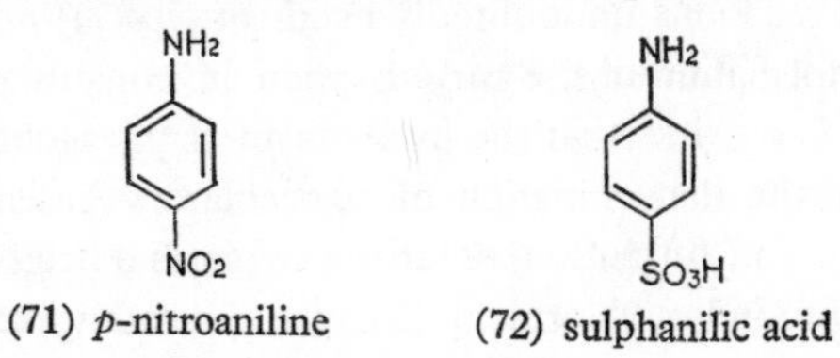

(71) *p*-nitroaniline (72) sulphanilic acid

diazotised benzidine (diaminodiphenyl, $NH_2—C_6H_4—C_6H_4—NH_2$) are most used. The different phenolic acids give different colours which change characteristically with pH. Thus after spraying with the diazo reagent, the papers are sprayed with aqueous Na_2CO_3 (4.6).

2.9. *Condensation reactions with aldehydes*

The enhancement of the reactivity of the carbon atoms in *ortho* and *para* positions to a phenolic hydroxyl group allows substitution reactions to take place which are not possible with aromatic hydrocarbons. One of these reactions is the condensation of phenols with aldehydes in acid solution.

In the presence of strong acids, carbonyl compounds especially aldehydes (73) acquire a proton to form an electrophilic radical (74), which is able to attack the aromatic nucleus of a phenol (75) to give a compound (76).

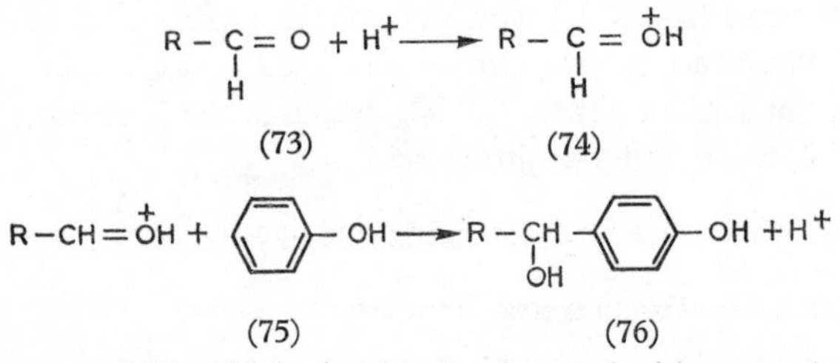

In the case of formaldehyde, the ring is attacked in several positions with the removal of water, resulting in the formation of polymers (77) of high molecular weight—phenol-formaldehyde resins—which are used in the manufacture of various plastics (e.g. bakelite). The phenolic compounds present in a plant extract can similarly be removed, by precipitating them with formaldehyde in acid solution.

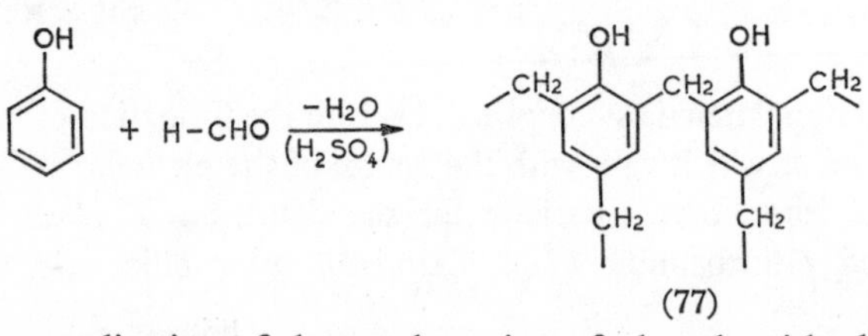

Another application of the condensation of phenols with aldehydes is in the use of vanillin (vanillaldehyde) as a spray reagent for chromatograms, or for the determination of particular natural phenolic compounds. Vanillin (78), in acid solution, captures a proton and gives an electrophilic radical (79) which, like the preceding one, is able to condense with the aromatic nucleus. This radical is, however, less electrophilic than that of formaldehyde because the mobile π electrons of the benzene ring of vanillin tend to fill the gap in the electronic environment of the $=\overset{+}{O}H$ group and thereby decrease its (positive) electrophilic character. Finally, this radical reacts only on those benzene rings which are 'activated', namely resorcinol or phloroglucinol nuclei (2.7).

Thus, vanillin (82) reacts with the A ring of leucocyanidin (80) (5,7,3′,4′-tetrahydroxyflavan-3,4-diol) at one of positions 6 or 8 activated by the presence of several OH groups. It does not react with the B ring. In the case of flavones or flavonols, which have a CO group in position 4 (70), the A ring is 'deactivated' and does not react, nor, as in the previous case, does the B ring react. In the case of leucocyanidin (80), the radical (79), resulting from the protonation of the aldehyde (78), is substituted in positions 6 and 8 exclusively, to give an intermediate (81) which is readily dehydrated to (82), the end product of the reaction (Swain and Goldstein, 1964). This is red in colour and is useful both in the development of paper chromatograms and in analysis (7.13).

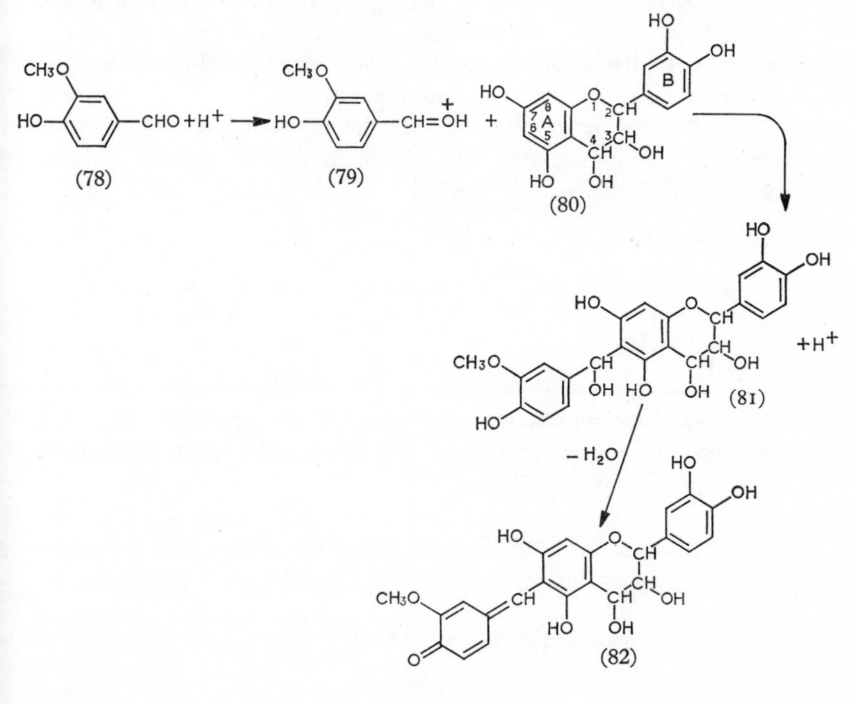

2.10. *Various reactions*

In this section several reactions are described which have been used in the study of natural extracts (Swain and Goldstein, 1964).

Nitrosation is the substitution of a hydrogen atom of a benzene ring with the radical NO_2^+. It requires an aromatic nucleus sufficiently activated, as is the case with phenols in which the *ortho* and *para* positions are free. The action of sodium nitrite in acid solution on phenol (83) yields *p*-nitrosophenol (84), which when made alkaline gives the yellow-coloured benzoquinone oxime (85), λ_{max} 420 nm.

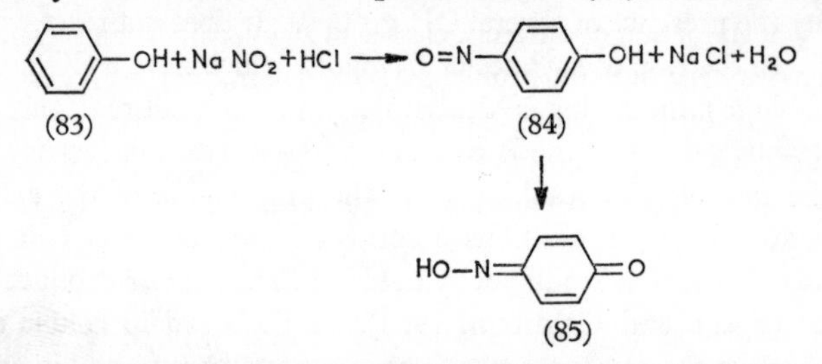

Hoepfner's reaction is an example of a nitrosation reaction. It was described for the detection of chlorogenic acid, which gives a bright red colour, but many other phenolic compounds react to give a coloration close to that of chlorogenic acid. Bate-Smith (1956b) has described precise conditions for carrying out the test.

Gibbs' reagent makes use of the reaction between 2,6-dibromobenzoquinone chloroimine (86) and phenol in alkaline solution to give 2,6-dibromoindophenol (87), which is blue in colour (λ_{max} 620 nm).

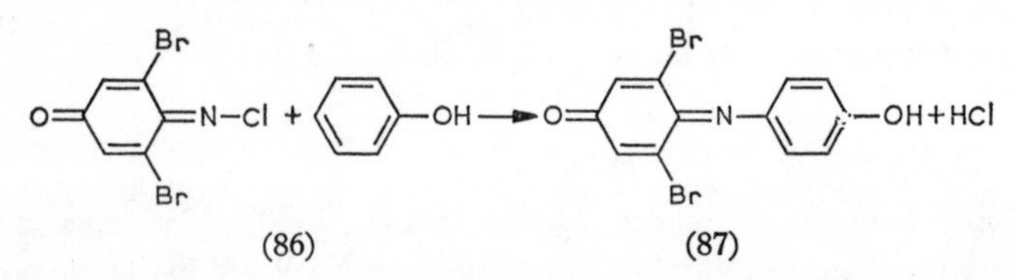

Lastly, 4-aminophenazone (88) gives coloured products (89) with phenols, in the presence of alkali and an oxidising agent, which have λ_{max} between 500 and 520 nm.

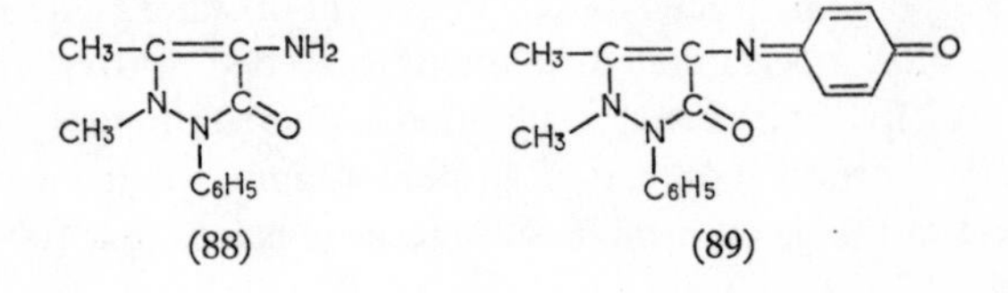

Use of the properties of phenols in their determination in natural products

2.11. *General considerations*

In this section, the two main methods for the determination of total phenols, the Folin-Denis and the permanganate methods (2.6) will be described. The possibility will then be discussed of determining separately the different types of phenolic function, in particular the *ortho*-diphenols and the *vic*-triphenols, using a method dependent upon complex formation with metals (2.4).

The determination of different classes of phenols in a complex natural mixture is a very difficult problem which has not yet been completely solved. Most often paper chromatography or differential spectrophotometry is used (Lebreton *et al.*, 1967). Paper chromatography provides a satisfactory separation of the various phenolic constituents of a complex mixture (Chapter 3) but the estimation of the substances so separated is not at all easy. In the case of spectrophotometry, substances absorbing at closely similar wavelengths are so often present that it is impossible to determine the concentration of any one constituent by a single measurement of absorbance. It is customary in such circumstances to measure the absorbance at one particular wavelength before and after treatment with a reagent specific for that particular constituent or for the group of substances to which it belongs.

Polarography can also be used for the determination of phenolic compounds, but this method is not yet as well developed as the preceding ones. Chouteau and Loche (1962) have used it for the determination of chlorogenic acid in tobacco.

The determination of tannins will be discussed in Chapter 7.

2.12. *Determination of total phenols*

Use of Folin-Denis reagent (2.6) (A.O.A.C., 1960) The reagent is prepared by mixing 750 ml of water, 100 g of Na_2WO_4, $2H_2O$, 20 g of phosphomolybdic acid and 50 ml of phosphoric acid, boiling under reflux for 2 hr and, after cooling, making up to 1 litre.

A solution of sodium carbonate is also prepared, by dissolving, at 70-80°, 35 g of anhydrous Na_2CO_3 in 100 ml of water. After cooling overnight, the supersaturated solution is seeded with a crystal of $Na_2CO_3 . 10\,H_2O$, and after crystallisation is filtered through glass wool.

To carry out the determination, an aliquot sample containing approximately 0·1 to 1·0 mg of phenolic substances is placed in a 100-ml flask containing about 75 ml of water, 5 ml of the Folin-Denis reagent and 10 ml of the Na_2CO_3 solution are added and the volume made up to 100 ml. After shaking, the absorbance is measured at 730-760 nm in a 1-cm cell. Swain and Hillis (1959) have modified the method to apply to smaller volumes.

It is often recommended that the result be expressed in terms of milligrams per litre of tannic acid, by comparison with a calibration curve prepared from a reference solution containing 100 mg of tannic acid per litre. This has little to be said in its favour, because tannic acid is so seldom present in plant extracts. It is preferable to express the results in terms of absorbance.

Use of potassium permanganate (2.6) The method to be described is that of Ribéreau-Gayon (J.) and Peynaud (1958) for the 'permanganate index' of red wine.

A 0·01 N solution of potassium permanganate is used, freshly made up from a 0·1 N solution. A solution of indigo carmine is also prepared, made up from 50 ml of a filtered stock solution containing 3 g per litre, 50 ml of concentrated sulphuric acid and 100 ml of distilled water.

The titration of the phenolic constituents of wine is carried out by measuring 50 ml of the above solution and 2 ml of wine containing 2-5 mg of phenolic substances into a beaker. $KMnO_4$ is added dropwise, increasingly slowly towards the end, until the blue colour completely disappears and is replaced by a clear yellow. Let n be the number of ml of $KMnO_4$ added.

To take into account the effect of the alcohol and organic acids present in the wine, 2 ml of a solution containing 5 g per litre of half-neutralised tartaric acid and 10% of alcohol are titrated under the same conditions. Let n' be the number of ml of $KMnO_4$ required.

The phenolic substances in 2 ml of wine will then have required for their oxidation $n-n'$ ml of $KMnO_4$; for 1 litre $5\,(n-n')$ ml of N $KMnO_4$ would have been required. The 'permanganate index' of the wine is $5\,(n-n')$.

2.13. *Separate determination of the different types of phenolic functions*

The method to be described for the determination of *o*-dihydroxy and *vic*-trihydroxy groups is that of King and White (1956) and is based on the work of several earlier authors, especially that of Kursanov and Zaprometov (1949).

This method depends upon the formation of a stable red-violet colour with ferrous salts in the presence of sodium potassium tartrate. In these circumstances, *m*-diphenols do not react, although they do react with ferric salts. The actual colour depends on the chemical composition of the medium and especially upon its pH; and the intensity of the colour is also different for catechol groups (*o*-diOH) and pyrogallol groups (*vic*-triOH).

King and White (1956) showed that at pH 8·3 the colour formed with pyrogallol in acetate and borate buffer is the same; but in the case of catechol the colour is twice as intense in acetate buffer (Table 3).

Table 3. *Standard values for the colorimetric determination of catechol and pyrogallol in the presence of ferrous salts, in two different buffer solutions at pH* 8·3 (absorbances: King and White 1956)

Weight (mg)	Pyrogallol Sample		Catechol Sample	
	Borate	Acetate	Borate	Acetate
5	0·977	0·994	0·366	0·707
4	0·787	0·797	0·294	0·577
3	0·597	0·597	0·225	0·435
2	0·390	0·388	0·157	0·294
1	0·199	0·194	0·082	0·147

To determine the *o*-diOH and the *vic*-triOH groups separately, therefore, in a mixture containing both, the reaction is carried out in acetate buffer and in borate buffer, respectively, both at pH 8·3, colorimetric determinations being made at 545 nm. The differences in absorbance between the two buffers are due solely to the contributions of the *o*-diOH groups. The concentration of these is calculated from a calibration curve obtained by titration of catechol, and is expressed in terms of catechol. The absorption corresponding to this concentration deducted from that of the sample, gives that due to the pyrogallol component. Table 3 gives an example of analytical values drawn from the paper by King and White (1956).

The reagent is prepared by dissolving 1·0 g of anhydrous ferrous sulphate and 5·0 g of sodium potassium tartrate in 1 litre of water, the solution being kept for 48 hr in the refrigerator. The borate buffer is prepared by dissolving 3·092 g of boric acid and 3·728 g of potassium chloride in 250 ml of water and adding enough N NaOH to bring the pH to

8·1-8·3. The acetate buffer is prepared from 10% aqueous ammonium acetate by adding at the time it is to be used, enough 10% NH_4OH to bring the pH to 8·1-8·3.

In carrying out the determination, a sample is taken containing 1 to 5 mg of phenols, to which are added 25 ml of the solution of ferrous tartrate and 10 ml of one of the buffer solutions. The volume is made up to 100 ml. (The buffer is added after the reagent in order to avoid precipitation.) The colorimetric measurements are made at 545 nm.

In applying the method to different substances (gallic acid and its derivatives,(+)-catechin), King and White (1956) found that the results conformed with the theoretical. They found, however, that if one of the hydroxyls in a *vic*-triOH compound is combined in ester form, the substance behaves as if it contained a pyrogallol group; whereas if it is combined as an ether, it behaves like catechol.

King and White (1956) also considered the possibility of determining *m*-diOH groups specifically, and especially the resorcinol nucleus. This method employs a reaction, due to Denigés, which consists in heating 10 mg of the phenolic compound with 2 ml of syrupy H_3PO_4 at 200-220°. After cooling and addition of 10 ml of water, 20 ml of 10% NH_4OH are added. If resorcinol is present, an orange colour is produced, with green fluorescence becoming bluish in ultraviolet light. Measurement of this fluorescence, or of the orange colour, might afford a means of determination, but other groups react with similar intensity, even though with different colours. Its lack of specificity therefore limits its possible usefulness.

2.14. *Determination of individual constituents. Use of chromatography*

Given the large number of different phenolic substances having such similar properties usually present in a natural extract, the determination of any particular one of them presents a difficult problem. Paper chromatography achieves a good separation of phenolic compounds, so that it is natural to seek to make use of this technique. It is not easy, however, to apply volumes of the order of a microlitre to a sheet of paper with any great accuracy. Furthermore, the analysis of the substances when separated is not an easy matter, for it entails elution followed by a determination, usually by visible or ultraviolet spectrophotometry, of a spot on paper which has first to be located. Elution of such microquantities is not easy to carry out quantitatively, and the solvents employed often extract impurities from the paper (3.16).

Elution can be carried out either by capillary displacement of the strip containing the desired constituent, or by immersing the paper containing it in a minimal quantity of solvent. In either case, the eluate is compared with a control eluate of an equal area of paper. The solvent most often used for elution is 70% aqueous ethanol, the case of anthocyanins being an exception since an acid solvent is required (methanol containing 0·1% HCl or 15% aqueous acetic acid).

If completely satisfactory elution cannot be attained, it is necessary to be content with visual comparison of the area and intensity of the spot with that of a standard control. Depending upon the nature of the phenolic compound, the spots are compared in visible light, ultraviolet light, or after treatment with a spray reagent. This method is a semi-quantitative estimation rather than a determination in a true sense. It is, however, often adequate in the absence of more rigorous methods. Greater sensitivity can sometimes be achieved by plotting the extinction of the chromatographic band, followed by planimetric measurement of the curve so obtained. The area under the peak is, under certain conditions, proportional to the concentration of the substance concerned. Various kinds of commercial apparatus are available to carry out this technique.

Whatever method is used, it is necessary, in order to determine the concentration of each substance, to compare the values obtained with those of known amounts of compounds. However, since these are often difficult to prepare, or to obtain by other means, it is necessary sometimes to be content with expressing the results in terms of the proportions of the different substances present. The results so obtained, although not having absolute analytical value, can be converted into indices and so facilitate comparisons between different samples. Such methods have been described, for instance, in the case of anthocyanins (Ribéreau-Gayon, 1959) and phenolic acids (Metche *et al.*, 1962).

2.15. *Determination of individual constituents. Use of spectroscopy*

Simple phenols absorb in the ultraviolet around 270 nm. In natural complexes, this peak may be displaced, and in addition one or (exceptionally) more peaks may appear, due to the presence of a chain of carbon atoms linked to one or more benzene rings. The method of measuring absorption spectra will be described in the next chapter (3.16).

Absorption in the ultraviolet has sometimes been used for the determination of total phenols in plant extracts, but this method is open to criticism. The results obtained with different extracts cannot properly be compared because of the very wide variation in the molecular absorption coefficient (ϵ) of even quite closely related phenols. For example, Vuataz *et al.* (1959) have shown that the values of ϵ vary between 1,285 and 3,120 for different catechins and between 11,500 and 13,500 for their esters with gallic acid.

In order to determine by spectrophotometry one particular substance, there is no difficulty provided it is in pure solution. It is necessary, however, to have a sample of the pure compound in order to be able to determine its absorption coefficient $\left(\epsilon = \frac{E}{CL}\right.$, where E = absorbance, C = concentration in g mol per litre, L = length of cell; it is more usual to employ log ϵ). The values of ϵ to be found in the literature are not very numerous. It must also be realised that the absorption of natural phenols is affected by many factors (nature of solvent, pH, etc.) so that it is necessary for each operator to determine standard values for himself under the conditions at which his observations will be made.

This change in the absorption with change in the nature of the solvent can be used in the determination of a single component, or a class of components in a complex mixture. Thus pH may be changed (see ionisation of phenols, 2.12) or metal complexes may be formed (2.4). Such methods consist in the measurement of absorption before and after the addition of a specific reagent for the particular substance or class. The change observed is referred to a calibration curve relevant to the substances to be determined. This method, known as the $\Delta\epsilon$ method, has been developed by, among others, Aulin-Erdtman (1957) and Mentzer and Jouanneteau (1957). Its use is, however, restricted to the solution of certain particular problems, because of the complexity of most natural extracts and the difficulty of finding a reaction completely specific for one definite group of substances. The method of determining *ortho*-di and *vic*-trihydroxy phenols is based on this principle.

Determination of anthocyanins There is no doubt that this method has found its widest application in the determination of anthocyanins. These are the only phenolic compounds possessing a red colour* (λ_{max} 520-550 nm). Generally, however, it is not possible to make a determination by direct spectrophotometry, because there are other substances present in plant extracts, especially tannins, which have a brown colour and interfere with the measurement. In the case of

* The betacyanins and anthraquinones must, however, be excepted from this generalisation. (Translator's note—E. C. B.-S.)

anthocyanins there are several reagents (6.4, 6.5) which modify their colour specifically and enable them to be determined, in particular, the decrease in colour with increase in pH (Sondheimer and Kertesz, 1948), and decolorisation by bisulphite (Dickinson and Gawler, 1956) or hydrogen peroxide (Swain and Hillis, 1959).

In using the method of difference in pH the absorbance is first measured (λ_{max} 520-550 nm) in buffer solution at pH 3·5-4·0, then after addition of 2% HCl at pH 0·5. (At this pH accuracy is not important to 0·1-0·2 pH.) The difference in absorbance is referred to a standard curve, prepared from a reference compound under similar conditions. This method requires that the sample may be diluted in order to buffer it to the required pH.

In carrying out the method based on decolorisation with hydrogen peroxide, Swain and Hillis (1959) use a reagent containing 1 ml 30% H_2O_2 and 9 ml of a mixture consisting of 5 parts (by volume) of methanol and 1 part of conc. HCl. These authors were using methanolic extracts of plant tissues containing 80-85% methanol and 0·5 N with respect to HCl. 3 ml of the extract are placed in each of two test tubes, 1 ml of the reagent is added to one of the tubes, and 1 ml of the methanolic HCl (5:1) to the other. After 5 min standing in the dark, the absorption of the two tubes is measured against that of water at 525 nm. As in the previous case, the difference in absorbance is referred to a standard curve.

In applying the method of decolorisation by bisulphite to red wines, Ribéreau-Gayon and Stonestreet (1965b) found it best to operate in an acid medium (2% HCl) in which the colour was more stable, but in these circumstances the sulphurous acid formed from the bisulphite is only weakly dissociated and the concentration of the SO_3H^- ion, which is the active reagent, is much decreased. It is necessary, therefore, to use a high concentration of bisulphite. To 1 ml of red wine are added 1 ml of the solvent used to dissolve the reference compound (1% ethanolic HCl) and 20 ml 2% HCl. 10 ml of this solution are placed in each of the two test tubes. To one are added 4 ml H_2O, to the other 4 ml of a 15% (approx.) solution of sodium bisulphite; having regard to the large excess of bisulphite its exact concentration is unimportant. The absorbances of the two solutions are measured at 520 nm, and the difference is referred to a standard curve.

The quantitative estimation of anthocyanins has also been studied in some detail by Fuleki and Francis (1968).

3

General Methods of Investigation of Phenolic Compounds

Introduction

Without doubt, paper chromatography has played a major part in advancing our knowledge in many areas of biological chemistry (Lederer and Lederer, 1957; Lederer, 1959, 1960). It is in the realm of phenolic compounds, and especially of the flavonoids, however, that the greatest benefits have been enjoyed, for the following reasons. First, they separate well on paper and only a limited number of solvents are required; second, they are easily revealed, in many instances on account of their own colour or their appearance in ultraviolet light; and lastly, because methods are available for identifying them in micro quantities on paper chromatograms.

Column chromatography and thin-layer chromatography (Randerath, 1964) are not so well developed in the case of phenolic compounds. Examples of the application of these methods will be given.

The tannins (Chapter 7) present their own particular problems. In their case it is necessary not only to determine the identity of the unit molecules of which they are composed, but also the manner and degree of their condensation. The same is also true of the lignins. In all other cases, paper chromatographic techniques are able to separate and identify the phenolic constituents in natural extracts.

It is necessary, however, to emphasise that phenolic compounds do not, as a general rule, exist in a free state in plant tissues. The exceptions are, first, the flavans, progenitors of the condensed tannins; and, second, substances included within the flavonoids but not, in fact, truly phenolic, for instance fully alkylated hydroxyflavones (Geissman, 1955). Free phenolic compounds are also found in certain reserve or dead tissues, for instance in seeds or heartwood. Except for these instances, aglycones in plant extracts are present only as a result of partial or complete hydrolysis during extraction.

Phenolic compounds occur in nature almost exclusively in the form of *O*-glycosides, forming bonds of the ether (acetal) type —C—O—C— (2.5). Sometimes, especially in the case of the phenolic acids, combination is in the form of esters, —CO—O—C—. Finally, there are some substances, e.g. vitexin, in which the combination with the sugar residue is carbon to carbon, —C—C—. These *C*-glycosides are resistant to hydrolysis and their investigation gives rise to particular problems (5.2).

In any one substance it is possible to have several different kinds of linkage simultaneously. For instance, in the anthocyanins both glycosidic and ester linkages may be present, the latter in the case of the acylated anthocyanins in which a cinnamic acid is esterified with a glycosidic sugar residue.

When starting work on unfamiliar plant material, it is always best first to identify the aglycones liberated by acid or alkaline hydrolysis. It is simplest to do this because, in any one tissue, a compound may be present in a variety of combined forms.

The term aglycone connotes the non-sugar moiety of a glycoside, i.e. the phenolic moiety. By extension, the term is sometimes used also for the phenolic moiety of esters, especially those of the phenolic acids.

The problems to be surmounted in the identification of the phenolic constituents of plant tissues are:

1. Extraction from the plant tissues.
2. Hydrolysis and separation of the aglycones.
3. Separation of the different combined forms.
4. Identification.

In writing this chapter, much use has been made of the many publications during the past few years on the subject (Harborne 1959a, 1959b; Seikel, 1962, 1964; Bate-Smith, 1964). The application of chromatography to the different types of phenolic compounds will be considered in greater detail in Chapters 4 and 7.

Extraction from plant material

3.1. *General principles*

The quantitative extraction of phenolic compounds from plant tissues is always a difficult problem and it does not seem that there is a completely satisfactory method. As an example, the author's experience with grape-skins in the course of anthocyanin investigations may be quoted

(Ribéreau-Gayon, 1959). The extraction was carried out with successive amounts of 1% aqueous hydrochloric acid, a solvent which gave the best results. An example of the results obtained is given in Table 4. The fourth fraction still contained 8% of the total anthocyanin extracted.

Although quantitative extraction is so difficult, the relative amounts of the different pigments in each fraction are reasonably constant, as can be seen in the lower half of the Table. The same result has been obtained with different glycosides (mono- and diglucosides of anthocyanins). It is possible, therefore, to obtain information about the relative amounts of the different phenolic compounds in plant tissues, even though they have not been extracted completely from the tissues. This is probably true in the case of the flavonoid glycosides, although it may not be necessarily true of all phenolics constituents, especially the tannins.

Table 4. *Successive extractions of the pigments of the skins* (10 *g dry wt*) *of Merlot grapes* (V. vinifera) *and the composition of the different fractions* (P. Ribéreau-Gayon, 1959)

Successive fractions	1	2	3	4
Volume (ml)	50	50	100	50
Absorbance (referred to 100 ml)	18·8	12·0	13·6	3·8
Percentages of the different pigments:				
Monoglucoside of:				
delphinidin	15	14	12	11
petunidin	16	13	13	14
malvidin	45	47	48	50
peonidin	13	12	12	13
acylated malvidin (2 pigments)	11	14	15	12

Another important aspect of the problem of extraction of the phenolic constituents is the possibility that changes may occur in the course of extraction. Plant cells contain many different enzymes liable to attack phenolic compounds, especially polyphenol oxidases and glycosidases. Under the action of the latter, complex glycosides may be degraded into simpler ones, or even into aglycones, which appear as artefacts on paper chromatograms.

As a general rule, these undesirable effects can be avoided by rapid drying of the plant material immediately after collection. Dried material can be preserved for some time without serious change. The material may also be frozen at −20° or −30°, at which temperatures enzyme action is completely inhibited; but it will reappear if and when the material is thawed for the purposes of extraction.

Examples of procedures for the extraction of various plant materials are to be found in the literature (Sannié and Sauvain, 1952; Geissman, 1955; Seshadri, 1962). Most often, the aim has been to isolate a constituent in pure crystalline condition. The development of paper chromatography has shown how difficult this task is, while at the same time rendering it less necessary, because it is so easy to identify a constituent by this means without having to isolate a weighable quantity of it. When a large number of phenolic compounds is present in the same tissues, it is possible to separate them fairly easily from the non-phenolic constituents (3.3) but very difficult to obtain each one of them in pure crystalline condition, because their properties are so similar. To obtain a pure compound in crystal form, it is necessary that the plant source should contain only a limited number of phenolic compounds. If this is not the case, purification processes are very difficult, if not in fact impossible.

3.2. *Solvents employed*

It is often recommended to make the first extraction of plant material with a non-polar solvent, such as light petroleum, which removes fats, waxes, chlorophyll and carotenoids more or less completely. If the extraction is not pushed too far, the flavonoids are not extracted by this solvent. There are exceptions: methylated flavones in leaves and fruits and heartwood flavonoids not glycosidically combined are often easily extracted by light petroleum (Geissman, 1955).

The extraction itself is carried out, after disintegration of the tissues in a blender, with methanol, ethanol or even with water. It is necessary to extract several times in order to achieve completely satisfactory results (cf. Table 4). Aqueous solvents have given the best results in the quantitative extraction of the anthocyanins of grapes. Methanol and ethanol have the advantage of being more easily removed, if it is desired to concentrate the solution *in vacuo*. If this is done, however, it is important not to proceed to complete dryness, because the residue is then difficult to dissolve. Water is to be preferred if fractionation is to follow by liquid-liquid extraction, for instance with ethyl acetate or ether, before or after hydrolysis.

Extraction of anthocyanins requires a weakly acid solvent (0·1% HCl is adequate) because these pigments are unstable in neutral or weakly alkaline solution (7.4). In adjusting the acidity, it is necessary to have in mind the presence of mineral constituents in the tissues which may increase pH. It is also necessary to take care not to hydrolyse flavonoid glycosides, some of which are easily hydrolysable, especially if the extraction is carried out at the boiling point of the solvent. For instance, if plant tissue is treated with boiling alcohol, the enzymes may be inactivated but the material is not protected against chemical hydrolysis by the action of the acids present in the tissue. Similarly, hydrolysis can occur in the course of a protracted extraction in apparatus of the Soxhlet type. Even though the temperature required to keep the solvent boiling may be low, it may be sufficient after a long period to bring about changes in the phenolic constituents. The author has abandoned the use of continuous extraction for the investigation of the phenolic acids and flavonols of grapes (Ribéreau-Gayon, 1964). If this method of extraction is employed, it is advisable to add sodium carbonate to the flask containing the boiling solvent in order to lower the acidity (Geissman, 1955).

3.3 *Purification of the extracts*

Having prepared an extract of the plant material, the next problem will be to ensure that it is suitable for chromatography. Often it will happen that the aqueous or alcoholic solution will be adequate for the purpose without further treatment, or with a simple extraction with a solvent such as ethyl acetate. However, many other constituents of plant tissues (e.g. sugars, acids, mineral salts) will be present and may interfere with chromatography, in which case purification will be necessary.

Bate-Smith (1964) has described a simple method which consists of boiling the plant material several times in a small quantity of water, followed by cooling, decanting, and combining the extracts. An equal volume of *n*-propanol is added, this being completely miscible with the water. When the mixture is saturated with sodium chloride, the propanol is 'salted out', separating as an upper phase, which contains, if not the totality of the phenolic glycosides, at any rate a representative sample of them. Since the moist propanol dissolves some sodium chloride, it is advisable to separate the supernatant layer, dry it in a current of warm air, and redissolve the phenolic substances in a non-aqueous solvent such as ethanol. The solution so obtained is very suitable for paper chromatography.

In the past, much use has been made of the ability of phenolic compounds to form insoluble precipitates with lead acetate under alkaline conditions, but this is not so much used at the present time. After centrifuging and washing, the precipitate is suspended in alcohol or water and decomposed with hydrogen sulphide. Alternatively, the lead can be removed and the phenols brought back into solution by treating with sulphuric acid or cationic ion exchangers. Up to a point it is possible, by treating successively with neutral and then with basic lead acetate, to separate some impurities and even to separate certain kinds of polyphenols, for instance *o*-diphenols and monophenols (Geissman, 1955). Such methods have, however, lost a good deal of their interest because chromatography achieves better results more simply and more certainly.

Cationic exchange resins are employed quite often for the purification of plant extracts containing phenolic compounds (Seikel, 1962). The resin most used is Amberlite IRC-50 in its H^+ form. The use has also been reported of Duolite Cation Selector CS-100; Dowex 1-XI, Amberlite 45 and Amberlite IR-4B. Except for the chalcone aglycones, phenolic substances in aqueous solution are absorbed by the resin. Mineral salts, sugars and acids can be removed by thorough washing with water. The phenolic glycosides are then eluted from the column with an aqueous alcohol, for instance 20% isopropanol, and the aglycones thereafter with 90% ethanol. Whilst this treatment achieves satisfactory purification of the original extract, it cannot provide true separation of the component phenols and it cannot, therefore, be used as a method of analysing them.

Swain and Nordström (1957) found it advantageous to remove sugars, which are often abundantly present in plant tissues, by the use of animal charcoal. This is often used in the reverse sense, to free sugar solutions from phenols which interfere in their determination. The phenols are, however, difficult to recover from the charcoal and a considerable proportion remain adsorbed irreversibly. Swain and Nordström resolved this problem by pretreating the charcoal with an aqueous solution of the same composition as that being investigated. When the charcoal is eluted with alcohol, a proportion of the flavonoids remains firmly attached, and when the experimental solution is applied and re-eluted, the phenols are recovered almost completely. The author prefers this method to the use of exchange resins because of the low adsorptive capacity of the latter.

Alternatively, it is sometimes possible to use a preparative pack of filter paper (3.14), by placing together several sheets of thick paper, preferably prewashed with water. If the extract is too complex, good separations cannot be obtained. These may however be good enough for a band to be cut off containing a particular constituent which, after elution, can be rechromatographed. Harborne (1959b) has used this method to isolate and identify the sugars in hydrolysates of glycosides. The solvents used were acetone-water (1 : 3), water and 15% acetic acid. In the first of these the sugars remain near the start-line; in the others they travel close to the solvent front.

It is sometimes possible to separate phenols by taking advantage of differences of acidity (2.2). To do this, they are transferred to a non-aqueous solvent (ether, butanol, ethyl acetate) from which they are extracted by aqueous solutions of gradually increasing alkalinity (sodium acetate, sodium bicarbonate, sodium carbonate). After acidification the aqueous solutions are re-extracted with ether, butanol or ethyl acetate. The most acidic phenols pass into the first, least alkaline, solvent. For instance, the phenolic acids are separated in this way from other phenolic constituents. Here again, the actual procedure must depend upon the particular situation.

Hydrolysis and separation of aglycones

3.4. *Acid hydrolysis*

Acid hydrolysis, which has the purpose of rupturing glycosidic linkages —C—O—C—, is carried out in N or 2N HCl, for a time which may vary from a few minutes to one hour, in a water bath at 100°C. The resistance of different glycosidic linkages to hydrolysis differs. In general, anthocyanins are more resistant than flavone and flavonol glycosides, but 7-glucosides and 7-glucuronides are particularly stable (5.10, Table 17). It is necessary to be aware of these facts, because glycosides and aglycones will appear together on chromatograms if hydrolysis is incomplete.

C-Glycosides (e.g. vitexin), in which the sugar residue and the flavone are linked through a carbon-carbon bond, are exceptionally resistant to both acid and enzymic hydrolysis, a property which can be made use of in their characterisation (Harborne, 1965a).

Besides the hydrolysis of glycosides, several other changes may take place, of which one should be aware, during heating with acid. The most important is the more or less partial conversion of leucoanthocyanidins (flavan-3,4-diols) into anthocyanidins (7.6). The same is true also of the condensed forms of these substances that are present in tannins. For this reason anthocyanidins present in acid hydrolysates may arise either from anthocyanins or from leucoanthocyanidins. Red colorations are often to be seen when tissues which do not contain anthocyanins are heated in acid solutions.

Along with the formation of anthocyanidins, the heating of leuco-anthocyanidins in acid gives rise also to polymerisation products (these are coloured red and are known as phlobaphenes). Catechins (flavan-3-ols) also give condensation products, but these are not coloured red. These polymers appear on chromatograms as ill-defined trails starting from the origin. The study of flavans, which are not naturally present as glycosides, should not, therefore, be carried out after acid hydrolysis (3.7).

Another change, which is not however so important, brought about in acid solution is that of flavanones into chalcones, accompanied by the formation of an intense yellow colour. If the 5 position of the flavanone carries a free OH group (1), the molecule is stabilised,

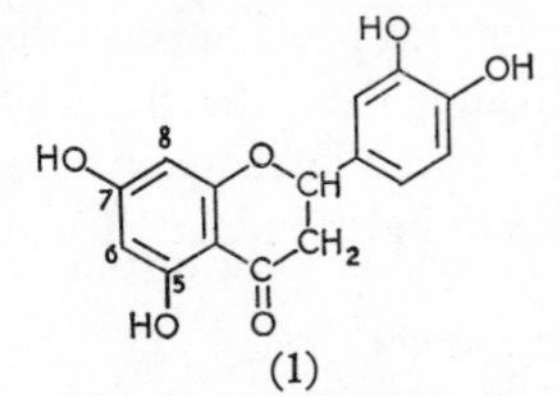

(1)

because a hydrogen bond is formed between this hydroxyl and the CO group. On the other hand, if this position is not hydroxylated (2), it is the isomeric chalcone which is more stable, because the hydrogen bond now forms between the CO group and the hydroxyl previously engaged in the heterocyclic ring (4) (Bate-Smith, 1964). This hydrogen bond is the more stable as it is involved in a ring of six atoms (2.3).

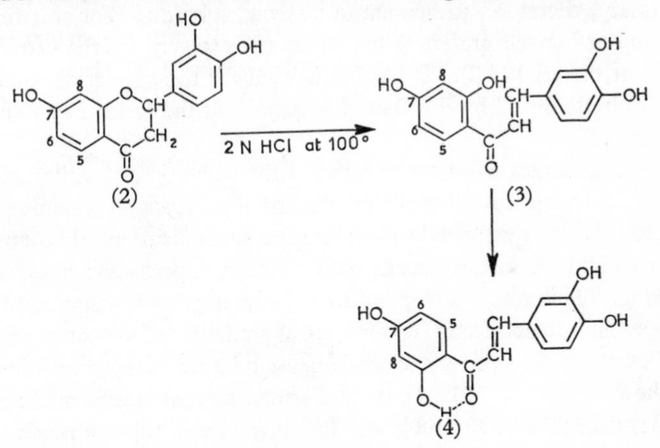

Enzymic hydrolysis can also be used to rupture glycosidic linkages. In this case the procedure is a little more complicated because particular enzymes can be used depending on the nature of the glycosidic sugar residue or the type of bond, and information can then be obtained on the nature of the pigment concerned (5.10, 6.15).

3.5. *Alkaline hydrolysis*

The alkaline hydrolysis of the esters of benzoic and cinnamic acid can be carried out at ordinary temperatures in 2N NaOH for 4 hours. It is best to work in an atmosphere of nitrogen, since many phenolic compounds are readily oxidised in alkaline solution (2.6). It is necessary to transfer to acid in order to examine the hydrolysis products.

This saponification is usually carried out in an aqueous extract of the plant material. In studying the phenolic constituents of wood, however, Pearl (1958) preferred boiling the material in N NaOH. In this way a large number of simple phenolic compounds are liberated, together with complex substances of indeterminate structure.

3.6. *Separation of the aglycones*

The procedure to be described comprises extraction of the plant material, followed by hydrolysis of two fractions of this extract, the one in acid, the other in alkali. In the case of the latter, the products are transferred to acid with the addition of HCl, preferably with cooling. The aglycones in the acid solutions so obtained are extracted with immiscible solvents (ether, ethyl acetate, isoamyl alcohol). It is necessary, therefore, to operate in aqueous solution. For this reason the author himself prefers to use an aqueous solvent for the extraction of plant material (3.2). If alcohol is used for the extraction, this has first to be removed by distillation *in vacuo* and the residue redissolved in water.

Some examples will now be given of the extraction of aglycones.

The solution resulting from alkaline hydrolysis (containing the phenolic acids) is extracted with ether after reacidification. The amount used should be just sufficient to produce a supernatant phase after shaking. This phase is applied to a chromatographic paper, and developed in butanol-acetic (upper layer of a mixture of *n*-butanol : acetic acid : water :: 4 : 1 : 5, or the single phase mixture of these constituents in the proportions 6 : 1 : 2); or in Forestal solvent (acetic acid : conc. HCl : water :: 300 : 30 : 100); or, for preference, toluene-acetic acid (upper layer of a mixture of toluene : acetic acid : water :: 4 : 1 : 5). The last solvent is especially recommended (4.5) because it gives a good separation of the phenolic acids which, in other solvents, do not separate well and become crowded at the solvent front. With toluene-acetic, it is necessary to equilibrate the paper for some hours with the aqueous phase of the solvent mixture before beginning the separation proper. The substances are revealed either by viewing in ultraviolet light or by spraying with diazotised nitroaniline (3.10).

The aglycones of the flavonoids are analysed in the solutions resulting from acid hydrolysis. These are treated differently according to the presence or absence in them of anthocyanidins, because these red pigments may mask the presence of other constituents:

1. In the absence of anthocyanidins, the hydrolysate is extracted once only with isoamyl alcohol, which extracts all the phenolic compounds present.
2. When anthocyanidins are present, all the other phenolic compounds are first removed by extraction with ether, then the anthocyanidins are extracted with isoamyl alcohol. In this case there are two extracts to be analysed.

As before, the solution is treated with just sufficient solvent to form a supernatant phase, which is then applied to the chromatographic paper. The Forestal mixture is the best solvent. Toluene-acetic acid may also be used, but butanol-acetic acid is useless for anthocyanidins because these require a more acid medium. For this purpose, therefore, butanol-HCl (upper layer of a mixture of *n*-butanol : 2N HCl :: 1 : 1) may be used.

Good separation of anthocyanidins, flavones and flavonols is achieved in Forestal, but many of the other phenolic constituents (benzoic and cinnamic acids and coumarins) have high R_f values and crowd together near the solvent front, where also are to be found any glycosides which may have escaped hydrolysis and brown oxidation products of some of the phenols. Chromatography in toluene-actetic acid gives a good separation of those constituents which have a high R_f value in Forestal solvent.

Ultimately if one-dimensional chromatography fails to effect a satisfactory separation, it may be advisable to run a two-dimensional chromatogram with butanol-acetic/Forestal of butanol-HCl/Forestal, when anthocyanidins are present.

Viewing in ultraviolet light is usually sufficient to locate the aglycones present on a chromatogram.

More complete details of these techniques are given in Chapters 4, 5 and 6. Just as an example, Fig. 1 shows the separation, after hydrolysis, of the phenolic aglycones of the skins of red grapes (*V. vinifera*). This figure also summarises the different procedures that have been described above.

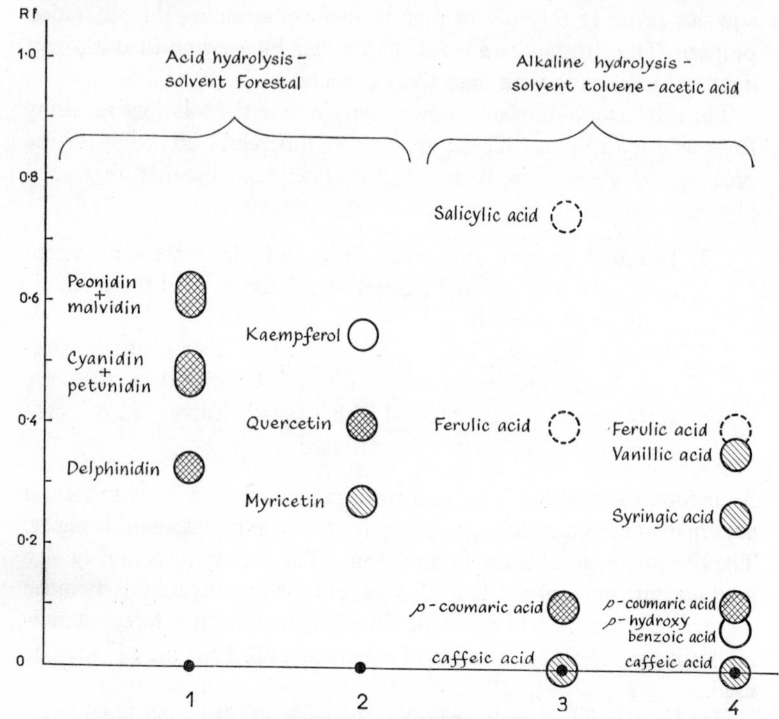

Fig. 1. *Chromatography of the simple phenolics of grape skins. 1. The red anthocyanidin spots mask the flavonols. 2. The flavonol spots are yellow in u.v. light; the cinnamic acids travel to the front. 3. The cinnamic acids and salicylic acid are blue fluorescent in u.v. light, changing colour with NH_3 (see 4.6, Table 9). 4. Same separation as 3.; but treated with* p-*nitroaniline and Na_2CO_3; each phenolic acid, except salicylic acid, gives a characteristic colour (see 4.6, Table 9)*

3.7. *The particular case of the flavans*

This item is introduced here because these substances behave differently from the other flavonoids. Their separation by paper chromatography will be discussed in detail in Chapter 7. This is an important question because the flavans are the precursors of the condensed tannins.

To separate and identify the flavans in the monomeric form (catechins and leucoanthocyanidins) it is obviously necessary to proceed without acid hydrolysis, because, for example, these substances do not occur in plant tissues in the form of glycosides, and also, they are changed by heating in acid solution (3.4). From aqueous solution, they are extracted with ethyl acetate, and after concentration they are separated by paper chromatography using butanol-acetic acid or dilute acetic acid (2 to 15%) as solvent. They are revealed either by acid vanillin reagent or by toluene-*p*-sulphonic acid (7.7, 7.8).

Separation of the different types of compound (glycosides and esters)

3.8. *Paper chromatography*

In this section, the separation of the different types of phenolic compounds will be considered, and more especially the separation of glycosides, in the forms in which they are present in plant tissues. The separation is carried out directly on the extract, without hydrolysis. It is advantageous to use ethanolic or methanolic extracts, because these can, if necessary, be concentrated more easily than aqueous extracts.

The basic solvent for the paper chromatography of these substances is the butanol-acetic acid previously described. If satisfactory separation is not achieved, water may be employed, using, if necessary, two-dimensional chromatography. In this case, to improve separation and avoid trailing, it is best to use a weakly acid solvent, usually acetic acid at a concentration between 2 and 15%. The author (Ribéreau-Gayon, 1953) was one of the first to use an aqueous solvent for the separation of anthocyanins, i.e. a solvent in which partition between two immiscible phases is not the only phenomenon concerned in the separation of the constituents. Factors of adsorption and differential solubility are also involved. The author used the lower layer of the butanol-acetic acid-water mixture, which is in effect an aqueous solution of acetic acid saturated with butanol. This solvent does not give precisely the same results as aqueous acetic acid. It is better for the separation of anthocyanins.

As Bate-Smith (1964) has pointed out, it is desirable to limit the number of solvents in common use to those whose properties are well characterised and always to report R_f values in these solvents. This facilitates a comparison of the results by different workers.

The solvents that have been described, Forestal and toluene acetic acid for the aglycones, butanol-acetic acid and water for the glycosides, are undoubtedly the most valuable and represent the basic solvents for the study of phenolic compounds. As things are at present, other solvents should not be used unless those mentioned above have proved inadequate for the solution of particular problems.

3.9. *Paper for chromatography*

Whatman No. 1 filter paper is the one most often used for the separation of flavonoids. It corresponds with the German paper Schleicher and Schüll 2043b MGL and the Japanese Toyo No. 50 (Harborne, 1959a). For preparative work, allowing an appreciable amount of a substance to be isolated, thick papers are used (Whatman No. 3 and 3 MM), which have a large absorptive capacity. The separations are slightly less satisfactory and it is sometimes necessary to complete purification on ordinary paper. In the case of anthocyanins, paper Arches 302 is recommended, since this gives better results than Whatman No. 1.

As regards the methodology of chromatography, there are of course several possible techniques (ascending, descending, circular, etc.) all of which have their own particular advantages and inconveniences. It has to be left to each operator to use the procedure best suited to his own problem. The author himself has always used descending chromatography.

3.10. *Identification of spots on chromatograms*

The ease of detection of phenolic compounds on paper is undoubtedly one of the reasons for the success of the use of chromatography in the investigation of these substances. Some flavonoids, for instance the anthocyanins, aurones and chalcones, are immediately evident because of their colour in visible light. A large number of other phenolic compounds are coloured when viewed in ultraviolet light, mostly in that of the Woods lamp (366 nm), but also at shorter wavelengths (253·7 nm). Some in fact are fluorescent. Luminescence is strongly intensified at low temperatures, for instance when the chromatogram is plunged into liquid nitrogen (−196°). Many of the benzoic acids, which are invisible at ordinary temperatures in ultraviolet, become intensely blue fluorescent when so treated (Ribéreau-Gayon, unpublished). In these circumstances some substances are phosphorescent, i.e. their luminescence persist when the exciting irradiation is discontinued.

In addition, phenols, on account of their acidic character, have different colours in acid and alkali (2.2). The physical properties of the phenolate ion are different from those of the free phenol, especially as regards colour in the ultraviolet. This property is much used in the examination of chromatograms in the vapour of conc. ammonia (sp. gr. 0·925, 20% NH_3).

Some substances, however (e.g. flavans), can only be revealed by spraying with an appropriate reagent. The principal tests which are used for the detection of phenolic compounds on chromatograms are set out in Table 5. Some glycosides give reactions which are different from those of their aglycones (5.6). There are, of course, other means of revealing constituents (Harborne, 1959b; Seikel, 1962, 1964). For instance, cinnamic acids and coumarins can be characterised with the aid of diazotised *p*-nitroaniline.

There are some reagents which react with all phenolic compounds. Their use, in spite of this lack of specificity, is often very helpful. In the first instance, use can be made of ammoniacal silver nitrate (2.6). A 5% solution of $AlCl_3$ in ethanol or a 5% aqueous solution of Na_2CO_3 can also be very useful, but in these cases the effect must be observed in ultraviolet light. The colours produced are different for the different classes of flavonoids. Lastly, a mixture in equal parts of 1% aqueous solutions of ferric chloride and potassium ferricyanide gives a blue colour with all phenolic substances.

3.11. *Column and thin layer chromatography*

Only a short section will be devoted to these subjects, because column chromatography has been much less applied to the investigation of phenolic compounds than has paper chromatography. It is worth recalling, however, that around 1935 Karrer and his associates used this method, little developed at that time, to separate the anthocyanins of some plants. These authors used a column of alumina on to which the mixture of pigments was applied in aqueous solution. The elution of the different substances was carried out with water, followed by dilute hydrochloric acid.

The use of cation exchange resins for the purification of plant extracts in order to obtain the total phenolic fraction has already been described (3.3). The separation of some catechins and flavonols has been achieved by Vancraenenbroeck *et al.* (1963) on columns of Amberlite CG.50 T1 or T2. Numerous other adsorbants have been tried, but few are really satisfactory. The most effective are magnesol (a synthetic commercial product consisting of a hydrated acid magnesium silicate) (Ice and Wender, 1952), polyamide powder, silicic acid, silica gel and cellulose powder.

Table 5. *Principal tests for the detection of phenolics on chromatograms*

Aglycone type	Colour in visible light	Colour		Special reagent	
		U.V.	U.V. + NH_3	Nature	Coloration
benzoic acid				diazotised *p*-nitroaniline (*)	different colours
cinnamic acid		blue	blue to blue-green		
coumarin		blue	blue		
anthocyanidin	red				
flavan				vanillin-HCl (**)	red
flavone	pale yellow	brown-black	bright yellow		
flavonol	pale yellow	bright yellow	bright yellow		
isoflavone		pale purple	pale purple	$AlCl_3$ in U.V. (***)	fluorescent yellow
flavanone			pale yellow	$AlCl_3$ in U.V. (***)	fluorescent green
aurone	bright yellow	bright yellow	bright red		
chalcone	yellow	brown-black	dark red		

* 2 ml *p*-nitroaniline 0·5% in 2N HCl, 3-5 parts 5% sodium nitrite, 8 ml 20% sodium acetate; after the first spray, the paper is then treated with 15% sodium carbonate to give further characteristic colours.

** 1 g vanillin dissolved in 10 ml conc. HCl.

*** Solution of 5% aluminium chloride in 95% ethanol; all classes of flavonoid react.

With the last, partition chromatography is carried out under the same conditions as chromatography on paper, and with the same solvents. Separation on the column allows appreciable quantities of the starting material to be dealt with. However, the same can be done with several layers of thick filter paper (Whatman No. 3 or 3 MM) and the separations are sharper.

In the case of alumina columns, the flavonoids are so strongly adsorbed that their elution is very difficult. In that of magnesol, the column is prepared by suspending the support material in dry acetone. The mixture to be separated is applied, in the same solvent, to the top of the column, which is developed first with aqueous ethyl acetate, then with aqueous ethanol, and finally with 95% ethanol. This method has been applied to numerous phenolic compounds.

Silicic acid, or silica gel, has been used to separate anthocyanins and other flavonoids. 10% aqueous phosphoric acid is first applied to the support as the fixed phase. Elution is effected by means of butanol or mixtures of butanol with benzene or chloroform.

Polyamide powder is, without doubt, the most successful medium for the application of column chromatography. This support has large adsorptive capacity and gives good separations. Furthermore, it is easy to use because it is only necessary to use water, or mixtures of water with alcohol or acetone, as solvent.

It has not been thought necessary to give experimental details of these various techniques, which are still little used, in view of the great superiority, in every respect, of paper chromatography. More detailed information is given by Seikel (1962, 1964), who remarks that it is generally possible to separate the different classes of flavonoids by column chromatography, but that it is much more difficult to separate individual constituents of the same class.

As regards chromatography on plates, or thin-layer chromatography (TLC) (Randerath, 1964), this does not seem likely to replace, in present circumstances, paper chromatography as the dominant technique in the investigation of phenolic compounds. It can, however, serve a valuable purpose (Nybom, 1964) because it is rapid and readily adapted to the serial analysis of large numbers of samples. It is coming increasingly into use. It has been successful in solving particular problems of identification and an example of this will be given later (5.6, 5.10). Polyamide powders and silica gels have been most used as supports for this purpose.

A complete chemical analysis of plant organs by TLC has been described by Nybom (Balsgård Fruit Breeding Institute, Fjälkestad, Sweden). The phenolic compounds are separated on a layer of cellulose powder MN 300 (Machery, Nagel and Co.) in two dimensions, with the solvents:

First direction	2% aqueous formic acid
Second direction	*n*-pentanol : acetic acid : water :: 20 : 12 : 10

Other solvents have been suggested by the same author (6.11) for separating anthocyanidins and anthocyanins.

To sum up, TLC has the versatility of column chromatography and the accuracy of separation of paper chromatography, as well as speed of operation, but the reproducibility of the R_f values is not quite as good. The spots can be detected in the same way as on paper, and in addition spray reagents such as concentrated acids can be used which paper cannot withstand.

It is also possible to separate phenolic compounds by gas liquid chromatography, after converting them into volatile derivatives such as their silyl ethers (Henglein and Kramer, 1959) or methyl esters (Harkins, 1965).

Identification

3.12. *General*

Up to this point, we have considered the operations involved in the separation by paper chromatography of the simple phenolics (aglycones) liberated by hydrolysis and the glycosides and other complex forms in which they are actually present in the plant tissues.

The identification of the aglycones is relatively easy. Various colour reactions (Table 5) allow an unknown substance to be allocated to a particular class. The number of substances per class being limited, identification by comparison of the R_f values with those of reference compounds (3.13) usually presents little difficulty. By way of confirmation, the component can be eluted and its absorption spectrum determined (3.16).

It is the identification of glycosides and other derivatives (e.g. esters) of phenolic compounds, therefore, to which most attention needs to be given. However, the methods described for these substances in the following sections (3.13, 3.14, 3.15) will apply equally to the identification of the aglycones. For any one simple phenolic compound there may exist many combined forms, involving various other molecules and various kinds of bonds. To identify these forms, it is necessary to isolate each substance so as to be able to study its chemical, chromatographic and spectral properties.

3.13. *Comparison of* R_f *values with those of reference compounds*

R_f values are not readily reproducible, because they are affected by numerous factors (temperature, humidity, quality of paper, etc.) which are difficult to control. To obtain reproducible values, which can be regarded as physical constants, many precautions need to be taken which seriously complicate laboratory procedures. For this reason, the author himself has never sought to obtain reproducible R_f's.

In these circumstances, it is not to be expected that an investigator will observe R_f values identical with those quoted in the literature. It is necessary to run one or more reference substances simultaneously with the unknown substance and on the same paper. It is also well known that the use of an R_f value in only one solvent is not sufficient proof of identity. It is necessary to have equality in several solvents and further confirmation by other methods, physical or chemical.

Except for simple phenolic compounds (aglycones), direct comparison of R_f values is in any case limited by the availability of the necessary range of reference compounds. Some flavonoids and related compounds are supplied by chemical manufacturers. (It is always advisable to check their purity by chromatography.) Otherwise, as Harborne (1959b) points out, particular plant sources can be used to provide marker substances. 'Instant tea', for example, is a source of kaempferol, quercetin and myricetin; from parsley seeds apigenin and luteolin can be extracted; cornflower and red roses are good sources of cyanidin 3,5-diglucoside.

However, the R_f is of undeniable interest, because its value relative to that of related compounds is more informative than its absolute magnitude. For example suppose that according to the literature substance A has a higher R_f value than another substance B. It is required to identify substance X of related structure, B being available for comparison but not A. If X has an R_f value lower than B, X cannot be either A or B. This reasoning is valid, even though the R_f found for B differs from that given in the literature, provided that the same solvent has been used.

Such simple deductions are themselves quite useful, but it is possible sometimes to proceed to actual identifications, because a relation exists between R_f value and chemical structure which is especially applicable to the flavonoids. This question has been discussed especially by Bate-Smith and Westall (1950) and Bate-Smith (1964). These authors found that, for a given class of compounds, the function,

$$R_M = \log\left(\frac{1}{R_f} - 1\right)$$

varies linearly (in butanol-acetic acid) with the number of OH groups and sugar residues. Substances with irregular R_f values have particular structural irregularities.

The study of the chromatographic behaviour of the flavonoids, as a function of their structure, in an alcoholic (butanol) or aqueous solvent (Roberts *et al.*, 1956; Harborne, 1959a, b; Ribéreau-Gayon, 1959) has led to the following conclusions which play an important part in the identification of these substances, especially having regard to the limited availability of the reference substances which are most often needed:

1. Increase in the number of hydroxyl groups lowers the R_f both in alcoholic (butanol-acetic) and aqueous solvents.
2. Methylation of hydroxyl groups increases the R_f in both solvents.
3. Glycosidation lowers the R_f in the alcoholic solvent but increases it in the aqueous solvent. The position of the sugar residue can have an important effect on R_f; for instance, flavonol glycosides with a free 3-OH group have zero R_f in water.
4. Acylation with a cinnamic acid increases R_f in both solvents.

In the case of the anthocyanins, Fig. 2 shows the variation of R_f as a function of chemical structure.

The importance of comparisons of R_f value for the identification of anthocyanins is well illustrated by Fig. 3, taken from the author's work on the anthocyanins of grapes (Ribéreau-Gayon, 1959). This figure represents schematically a two-dimensional chromatogram (solvents 1 and 2 being the upper and lower phases of a butanol-acetic acid-water mixture). It shows the positions of all the anthocyanin pigments met with in different varieties of grapes. A key to their identity is given in Table 6.

These pigments originate from four different kinds of combination of each of five anthocyanidins. For each kind of combination, the relative positions of the different aglycones is the same. A chromatographic chart of this kind is an invaluable guide to the identification of the pigments. This diagram also shows the value of two-dimensional

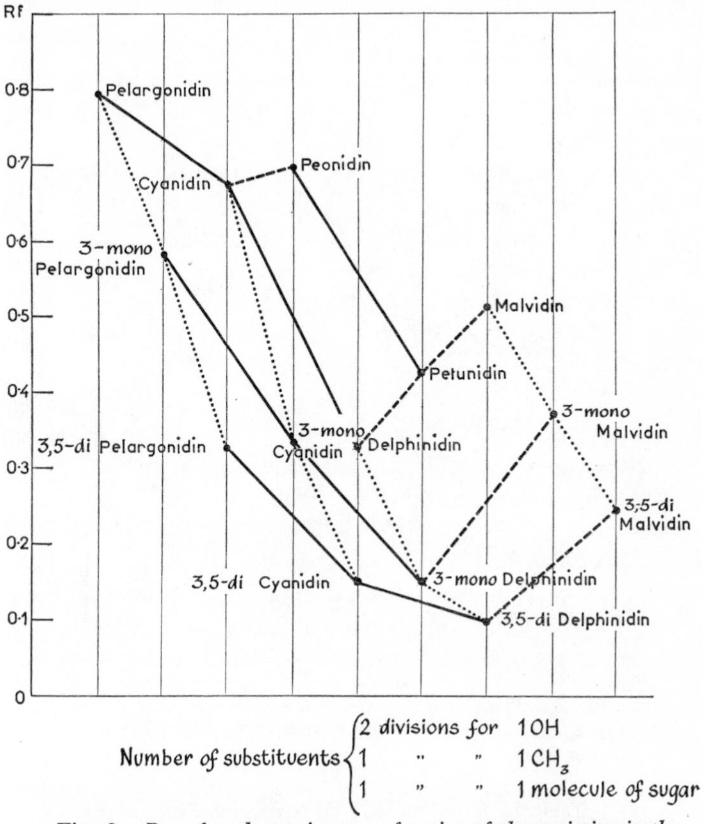

Fig. 2. *R_fs of anthocyanins as a function of the variation in the number of substituents, based on the data of Bate-Smith and Westall (1950)*

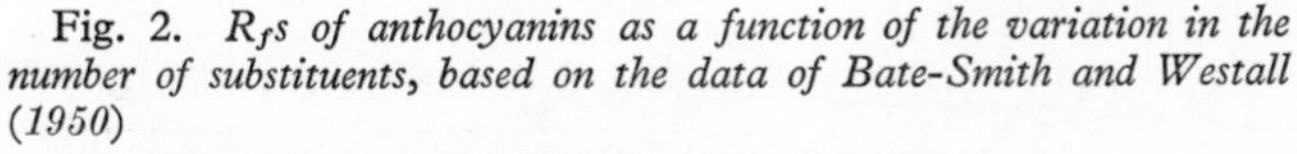

chromatography, and the impossibility of replacing it by one-dimensional chromatography in a number of solvents.

However valuable these methods are, they are still not sufficient for rigorous unambiguous identification. For this, the isolation of the constituent in a pure state, and its identification by chemical, spectrometric and chromatographic tests, is indispensable.

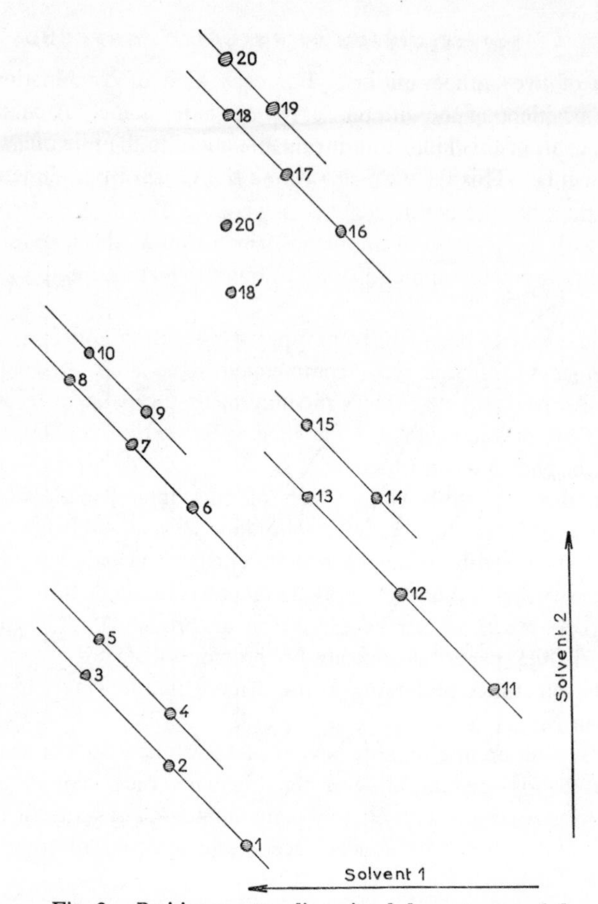

Fig. 3. *Position on a two dimensional chromatogram of the anthocyanins of grapes (solvents 1 and 2, cf. text) (P. Ribéreau-Gayon, 1959)*

Table 6. *Identification of the pigments in Figure* 3

Glycosides	Aglycones				
	Delphi-nidin	Petu-nidin	Malvi-din	Cyani-din	Peoni-din
Diglucoside	1	2	3	4	5
Acylated Diglucoside	6	7	8	9	10
Monoglucoside	11	12	13	14	15
Acylated Monoglucoside (1)	16	17	18	19	20
Acylated Monoglucoside (2)			18′		20′

(1) and (2). The difference in structure between these two pigments is not known.

3.14. *Isolation of pure products*

In this section the preparation in a pure state of a simple phenolic compound from a complex mixture, with a view to its ultimate identification, will be discussed. This objective is different from the purification of the phenolic constituents as a whole (3.3) which is concerned merely with the removal of impurities which would otherwise interfere with their chromatographic separation, but the methods employed are similar.

Recourse may be had to different types of chromatography, on column or on paper, but the method recommended is the latter, especially the preparative pack, because this is the only method capable of solving all the problems of fractionation, even when a plant extract contains a large number of phenolic constituents.

When this is possible, large sheets of thick filter paper (Whatman No. 3 or 3 MM) are employed, the solution to be analysed being applied over the whole width of the paper. On ten sheets it has been possible, by successive applications, to apply 25 ml of a solution of anthocyanins, which is as much as can be placed on a chromatographic column. However, thick papers do not always provide satisfactory separations and it is sometimes preferable to use a more free-flowing paper (e.g. Whatman No. 1).

After separation in a suitable solvent (usually butanol-acetic acid), the different bands are cut off and the pigments they contain eluted. Elution is carried out with ethanol containing 30% of water or, in the case of anthocyanins, with aqueous acetic acid at a concentration which may vary between 2 and 15%.

Figs. 4 and 5 are concerned with the same experiment. They show an example of chromatographic separation on thick paper of the anthocyanins of the grape (variety Seyve-Villard 18.315; Ribéreau-Gayon, 1959). The mixture is obviously a complex one, and there can be no question of isolating in one operation all the pigments corresponding to the eight bands A to H, which overlap to a greater or less extent.

Furthermore, some of the bands contain more than one pigment. This is demonstrated in Fig. 5, which shows, at the right-hand side (i), the schema of the one-dimensional chromatogram in Fig. 4, and at the left (ii), the corresponding result of chromatography in two dimensions. The horizontal lines indicate the levels at which the bands were cut. It can be seen that band D contains three pigments, D_1, D_2 and D_3, and band E two pigments, E_1 and E_2.

It is necessary, therefore, after having made a preliminary separation,

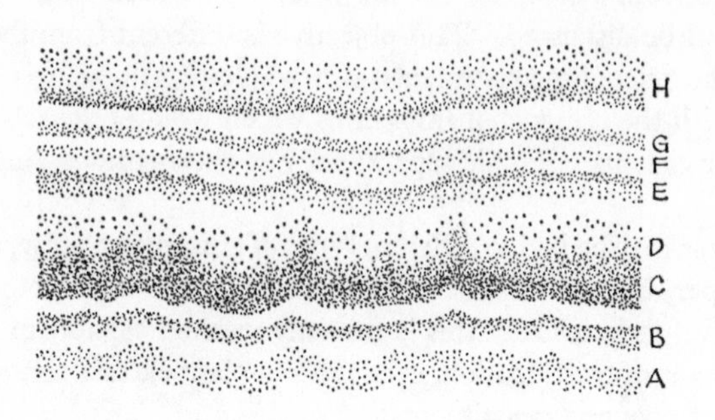

Fig. 4. *Example of the separation of grape anthocyanins in butanol-acetic on a large sheet of Whatman No 3 paper (A – H, cf. Fig. 5) (P. Ribéreau-Gayon, 1959)*

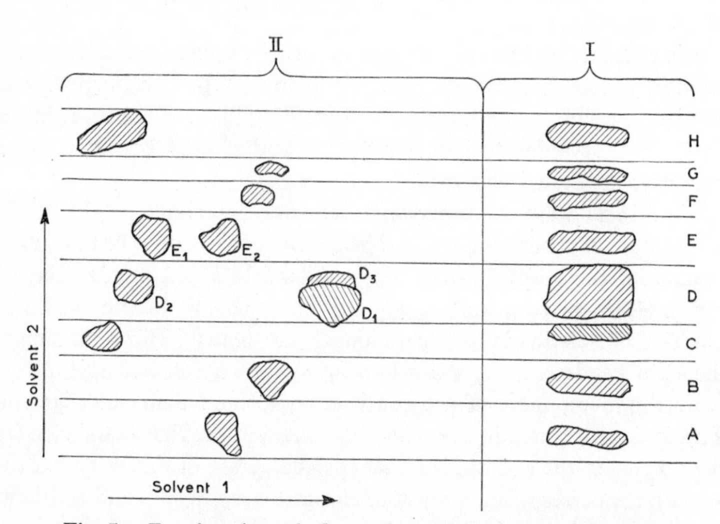

Fig. 5. *Fractionation of the anthocyanins of the grape Seyve-Villard 18.315 (solvents 1 and 2, cf. text) (P. Ribéreau-Gayon, 1959). I. Different fractions separated by one-dimensional chromatography in butanol-acetic (solvent 2). II. Correspondence with the two-dimensional chromatogram*

to repurify some of the fractions in a different solvent (dilute acetic acid). It may even be necessary to repeat the operation several times. In this way, each substance is eventually obtained in a pure state, but the operation often requires great care and is very time-consuming. It is always important to check the progress of the separations constantly by two-dimensional chromatography.

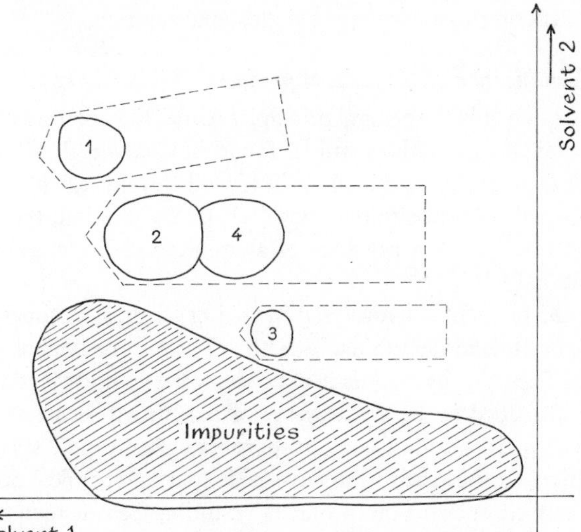

Fig. 6. *Two-dimensional chromatography of the flavone derivatives of grapes (the method for cutting out the different pigments is indicated by dotted lines; solvents 1 and 2, cf. text) (P. Ribéreau-Gayon, 1964a)*

Sometimes one-dimensional thick paper chromatography does not give sufficiently complete separations. An instance of this occurred during an investigation of the flavonol glycosides of grapes (Ribéreau-Gayon *et al.*, 1964a). It was found necessary to abandon thick paper because this produced trails which were quite inseparable, and use the thin Whatman 1 paper. Even so, it was necessary to cut out the fractions from a two-dimensional chromatogram, because the bands were too indeterminate on a one-dimensional paper. The result is shown in Fig. 6, solvent 1 being butanol-acetic acid, solvent 2 dilute acetic acid. Pigments 2 and 4 were later purified by a second separation in solvent 1. The chromatography was carried out simultaneously on ten papers. With the small amounts of substances so obtained, it was possible to identify all four pigments (3-monoglucosides of kaempferol (1), quercetin (2) and myricetin (3) and the 3-monoglucuronide of quercetin (4)).

Usually, the chromatographically pure substances so obtained are used directly in the eluate, but they can be crystallised by concentrating *in vacuo* and adding an appropriate second solvent. In the case of anthocyanins, precipitation is effected by addition of a large excess of anhydrous ether to a concentrated alcoholic solution.

3.15. *Identification of the isolated products*

In this section only the general principles of identification are considered. More detailed explanations will be found in Chapters 4, 5 and 6.

So far as possible, the R_f values in several solvents are first compared with those of reference compounds (3.13). In addition, the spectra in acid and alkali, and in presence of aluminium chloride, give essential information (3.16).

Next the pigment is hydrolysed in acid or alkali, sometimes, as in the case of acylated anthocyanins, in both. The simple phenol (aglycone) resulting from this hydrolysis is next identified (it must be one of those already identified in the hydrolysate of the initial extract, 3.6). This identification is accomplished by chromatography and spectroscopy. In addition, in the case of the flavonoids, partial chemical degradation can be carried out and the products resulting from it identified (6.5). The sugars and other residues involved in the structure of the natural product are similarly identified by chromatographic methods.

Finally, the mode of combination of the component parts of the complex molecule has to be determined. This is a difficult problem and there is no one method of universal application. The principal possibilities are discussed in the following chapters.

3.16. *Measurement of absorption spectra*

Spectroscopy in the visible and ultraviolet plays an important part in the identification of phenolic compounds. In spite of its undeniable possibilities, infrared spectroscopy is less used, because these substances are as a rule insoluble in transparent solvents (e.g. chloroform), so that special techniques (use of KBr discs) are necessary requiring the availability of crystalline products (5.13).

The preparation of the sample is the first problem encountered after the compound has been separated by paper chromatography. Bradfield and Flood (1952) described a method enabling the substance to be examined without removal from the paper. However, elution, which allows the operation to be carried out in solution, is the method most commonly used.

The most important problem here is the elimination of interfering substances that the solvent is likely to extract from the paper. Preliminary washing of the paper is seldom sufficient for this purpose, because further impurities are removed every time a solvent percolates through the paper. A way to combat this is to treat an equivalent piece of the paper in exactly the same way as the chromatogram itself, and with the solution so obtained cancel out the absorption of the impurities present in the experimental eluate.

For the same reason, care should be taken to prepare an eluate as concentrated as possible, so that it can be diluted with pure solvent. Since the impurities dissolved are proportional to the volume of solvent which has travelled over a given surface of paper, their concentration in the solution to be spectroscopically examined will be reduced to a minimum. If, for instance, one has a strip 20 cm long carrying the experimental substance, the following procedure may be used. As a first step, the solvent is allowed to run just to the end of the strip, so that the substance is concentrated in this region. The paper is then dried and the solvent allowed to run again, this time eluting practically the whole of the substance in the first few drops that drain from the end of the strip. These few drops are then diluted to 3 ml (the minimum volume to fill the usual 1 cm cell of the spectrophotometer) a second cell being filled with a solution prepared in exactly the same way from a blank strip of paper. Operating in this way, a vastly superior spectrum is obtained to one which would be obtained by eluting the paper with 3 ml of solvent. In this case the solution would contain much more impurity and it would not be possible to compensate it so rigorously with a control solution obtained under the same conditions.

Elution is usually carried out with ethanol containing 30% of water or in the case of anthocyanins with dilute (2 to 15%) acetic acid. The spectra may be traced after dilution with water or, more often, with an alcoholic solvent (ethanol or methanol). In the case of anthocyanins the solvent must be weakly acid, containing, for example, 0·01% HCl.

It is important to define carefully the nature of the solvent used and to compare the spectra of the substances under examination with those of reference substances under the same conditions. In the case of anthocyanins, the position of the absorption maximum in the visible region can vary 20-30 nm in changing from an alcoholic to an aqueous medium.

Much use is made, both in solving problems of identification and in quantitative analysis (2.15) of the shift in absorption with the addition of certain reagents. On addition of alkali, for instance, the spectrum is that of the phenolate ion, which differs from that of the free phenol; and a shift occurs with aluminium chloride as a result of complex formation (2.4). Other reagents employed are sodium acetate, sodium acetate-boric acid, magnesium sulphate, zirconium perchlorate, etc. (Mentzer and Jouanneteau, 1956; Jurd, 1962a; Harborne, 1964).

When operating in alcoholic solution, sodium ethylate is added to produce the alkaline shift; a 0·002 M solution is used; which may be prepared by adding 0·5 ml of 0·012 M sodium ethylate (itself prepared by dissolving about 0·3 g of sodium in 1 litre of ethanol) to 2·5 ml of an alcoholic solution of the pigment. Needless to say, the spectrum is measured against that of a reference solution prepared under the same conditions.

Aluminium chloride is added so as to produce an alcoholic solution containing 0·1% $AlCl_3$, by adding 0·5 ml of a 0·6% ethanolic solution to 2·5 ml of pigment solution, again using a reference solution in measuring the spectrum.

In the case of sodium acetate, an excess of solid is added directly to the pigment solution in the cell. This reagent is useful in detecting, under certain conditions, the presence of a free OH group in the 7 position of flavonoids (5.13).

To identify *o*-dihydroxy groups, Jurd (1956) uses the boric acid-sodium acetate mixture which produces a bathochromic shift (towards the longer wavelength) (5.11) in the presence of that grouping. To 8 ml of the alcoholic solution of the substance to be analysed are added 2 ml of saturated ethanolic boric acid and an excess of anhydrous sodium acetate. As an alternative technique, Markham and Mabry (1968b) add, to 2·5 ml of the pigment solution, 3 drops of a 5% methanolic solution of $AlCl_3$, then 3 drops of a solution containing 50 ml of conc. HCl and 50 ml H_2O. The difference between the spectra before and after addition of HCl enables *ortho*-dihydroxy groups to be detected (2.4, 5.7).

4

The Phenolic Acids and their Derivatives

Introduction

This chapter covers the benzoic acids, with seven carbon atoms (C_6—C_1), the cinnamic acids with nine (C_6—C_3) and the coumarins, which possess an oxygen heterocycle of six atoms fused with a benzene ring and are derived from cinnamic acids by cyclisation of their side chain. This is not the customary form of presentation in that the chemical and biological relationships between the three groups are not immediately evident.

The cinnamic acids are undoubtedly the best known of the constituents considered in this chapter; their distribution in the plant kingdom is very wide. However, a number of recent publications have indicated the frequent occurrence of several different acids derived from benzoic acid and it seems useful therefore to give them more prominence than is usual in reviews of this subject. Thus, for example, in Table 2 (1.1) certain benzoic acids are included among the principal phenolic constituents of plants, this table being based on one appearing in the review paper of Swain and Bate-Smith (1962) which does not, however, include the benzoic acids.

Several reasons have prompted us to consider the two types of phenolic compound together. In the first place they are both acidic and occur in plant tissues in the form of esters; it has been suggested, moreover, that the benzoic acids arise from the cinnamic acids, although this has never been definitely proved. Secondly, it is possible to separate them and identify them simultaneously on the same chromatographic paper after alkaline hydrolysis of the plant extract.

The coumarins are much less widespread than the cinnamic acids. It will not be possible to deal with them in great detail here, because of lack of space.

The nature of the substances and their distribution in the plant kingdom

4.1. *The benzoic acids*

Harborne and Simmonds (1964) pointed out that the benzoic acids represented in the structures (1)-(7) are widely distributed both in angiosperms and gymnosperms. Their distribution has been studied by Griffiths (1959), Ibrahim and Towers (1960), Tomaszewski (1960) and Ibrahim *et al.* (1962). Tomaszewski, for instance (1960) has identified *p*-hydroxybenzoic and gentisic acids in the leaves of 97% of plants sampled from 86 different families.

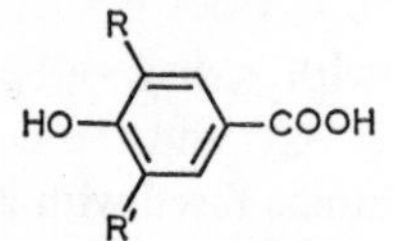

(1) R=R'=H; *p*-hydroxybenzoic acid
(2) R=OH, R'=H; protocatechuic acid
(3) R=OCH_3, R'=H; vanillic acid
(4) R=R'=OH; gallic acid
(5) R=R'=OCH_3; syringic acid

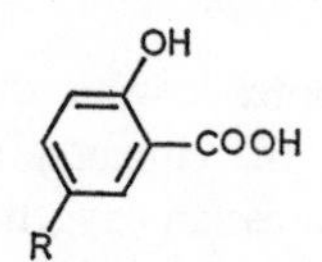

(6) R=H; salicylic acid (*o*-hydroxybenzoic acid)
(7) R=OH; gentisic acid

Gentisic acid (7) has a peculiar pattern of hydroxylation not found in any of the commoner flavonoids. It can be related to that of salicylic acid (6) but is, however, very much less common.

p-Hydroxybenzoic (1), vanillic (3) and syringic (5) acids are constituents of lignin, from which they are liberated by alkaline hydrolysis. It can be said, as a fair approximation, that plants which do not contain lignin do not contain these acids either (Harborne and Simmonds, 1964).

Protocatechuic (2) and gallic (4) acids probably have different functions in the plant from those of the other benzoic acids. The former is very widely distributed, the latter is rarer, and often found in nature as its dimer, ellagic acid, which is the dilactone of hexahydroxydiphenic acid (7.3). Gallic and ellagic acids play an important part as constituents of many tannins (7.2, 7.3) from which they are liberated by acid hydrolysis. The distribution of ellagic acid in nature, after acid hydrolysis of plant extracts, has been studied by Bate-Smith (1956). It is present in 15 of the 40 orders of dicotyledons, but is absent from monocotyledons, gymnosperms and ferns.

Other acids of related structure are found, but only very occasionally; completely methoxylated acids, for instance, which are not properly phenolic, such as anisic acid (*p*-methoxybenzoic), veratric acid (3,4-dimethoxybenzoic) and eudesmic or trimethoxygallic acid (3,4,5-trimethoxybenzoic).

4.2. *The cinnamic acids*

Four cinnamic acids are well known and widely distributed in plants: *p*-coumaric (8), caffeic (9), ferulic (10) and sinapic (11). At least one of these is found in practically all higher plants. According to Bate-Smith (1962b), *p*-coumaric acid is the commonest of all phenolic constituents. These acids exist in plant tissues in various combined forms (4.3).

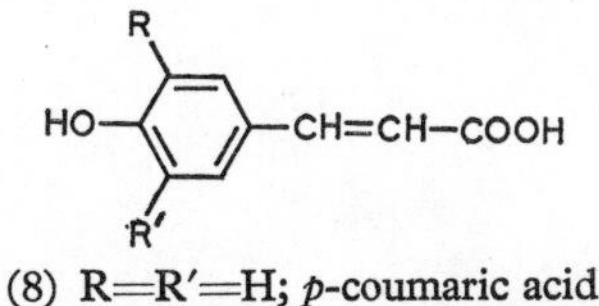

(8) R=R′=H; *p*-coumaric acid
(9) R=OH, R′=H; caffeic acid
(10) R=OCH_3, R′=H; ferulic acid
(11) R=R′=OCH_3; sinapic acid

It is essential to point out that the trihydroxycinnamic acid, corresponding to gallic acid, has never been found, which seems to indicate that it is in fact never formed. As Bate-Smith (1962b) points out, it is stable and would be easily identified on chromatograms if it were there.

This observation suggests that there must be some independence in the mechanism of synthesis of the different phenolic acids. This is supported by research with the acids in the skins of a variety of red grapes (Cabernet-Sauvignon) and one of white grapes (Sémillon) (Ribéreau-Gayon, 1964a). The results, given in Table 7, deal with total acids, i.e. analyses carried out after alkaline hydrolysis, which liberates the acids from the combination in which they are present.

There are other less common cinnamic acids. Cinnamic acid itself (12) has been reported in plants, but its detection is difficult because it does not fluoresce in the ultraviolet nor does it react with the classical phenolic reagents, such as diazotised *p*-nitroaniline.

o-Hydroxycinnamic acid (*o*-coumaric acid) (13) is to be regarded as uncommon. Since it has an OH group in the *ortho* position it cyclises very readily to give coumarin (14) which is a common odorous principle in plants.

Cinnamic acids, since they possess a double bond, are capable of existing in two isomeric forms, e.g. *cis*- (15) and *trans*-cinnamic acid (16).

As indicated by their structure, only the *cis* forms are able to cyclise to coumarins. The chemical synthesis of the *cis* and *trans* forms has been described by Ville *et al.* (1958) and by Courte *et al.* (1958). Different biological properties have been assigned to the *cis* and *trans* forms.

Table 7. *Phenolic acids in the skins of grapes* (Ribéreau-Gayon, 1964a)
(The results are expressed as mg per 1,000 grape skins. Dry wts per 1,000 grapes: Cabernet-Sauvignon, 62·7 g; Sémillon, 59·4 g. Dry wts of 1,000 fruits Cabernet-Sauvignon 1,090 g; Sémillon, 1,430 g.)

Phenolic Acids	Red Grapes (Cabernet-Sauvignon	White Grapes (Sémillon)
Cinnamic acids		
p-coumaric	12·0	0·5
ferulic	0·2	0·5
sinapic	0·0	0·0
Benzoic acids		
p-hydroxybenzoic	0·0	0·1
vanillic	3·0	0·1
syringic	12·0	traces

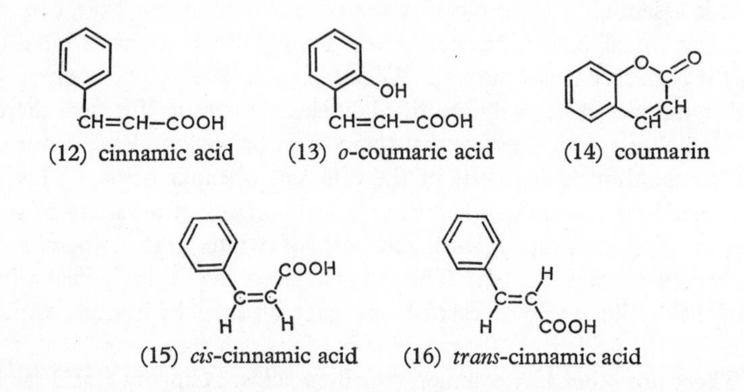

(12) cinnamic acid (13) *o*-coumaric acid (14) coumarin

(15) *cis*-cinnamic acid (16) *trans*-cinnamic acid

The naturally occurring cinnamic acids are the *trans* isomers, which are the more stable. However, the isomers are easily interconvertible under the action of light, an equilibrium condition being reached. This reaction has been studied by Kahnt (1967). The isomers are separable by chromatography in certain solvents such as dilute acetic acid (Williams, 1955), but not in butanol-acetic or toluene-acetic acid (*ibid.*). Similarly, the naturally occurring derivatives of the cinnamic acids, e.g. chlorogenic acid, exist as *cis* and *trans* isomers, which are separable by paper chromatography in aqueous solvents.

If a cinnamic acid, or a derivative thereof, is chromatographed after exposure for several hours to sunlight, using dilute acetic acid as a solvent, two spots having the same fluorescence in ultraviolet light and reacting similarly to phenolic reagents will be seen. If, after drying the paper, a second chromatogram is run in a direction at right angles, each of the spots will again appear as two. The *cis* and *trans* isomers have each undergone partial transformation into the other.

4.3. *Forms of combination of the phenolic acids*

Cinnamic acids The naturally occurring derivatives of the cinnamic acids, especially caffeic acid, have been known for a long time. They were first studied in coffee, which is particularly rich in these compounds (Pictet and Brandenberger, 1960) and from which caffeic acid derives its name.

Payen, in 1846, discovered a phenolic acid derivative in coffee, which on account of the green colour which it produced on standing with alkali, he named chlorogenic acid (17). Its structure was identified very much later. It is an ester of caffeic acid with a cyclic acid-alcohol, quinic acid; more precisely, it is 3-caffeoylquinic acid.

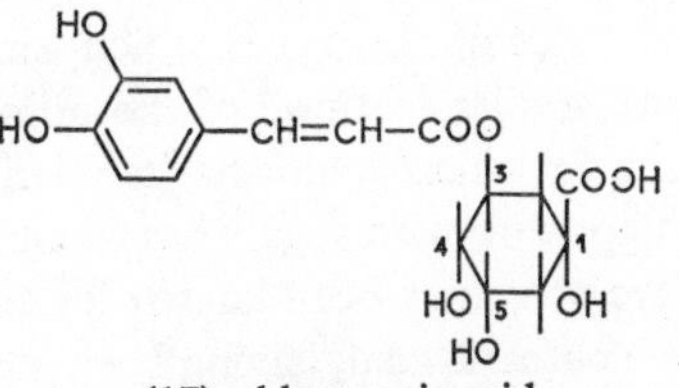

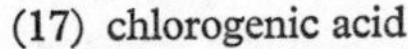

(17) chlorogenic acid

In 1950, Barnes *et al.* showed that coffee contains another compound of caffeic and quinic acids which they called isochlorogenic acid, the structure of which was difficult to elucidate. It is now known that it is a mixture of which the three main constituents are isomeric dicaffeoylquinic acids (Scarpati and Guiso, 1963; Corse *et al.*, 1965).

Yet other derivatives of caffeic and quinic acids are known. Neochlorogenic acid is 5-caffeoylquinic acid (18). A derivative designated

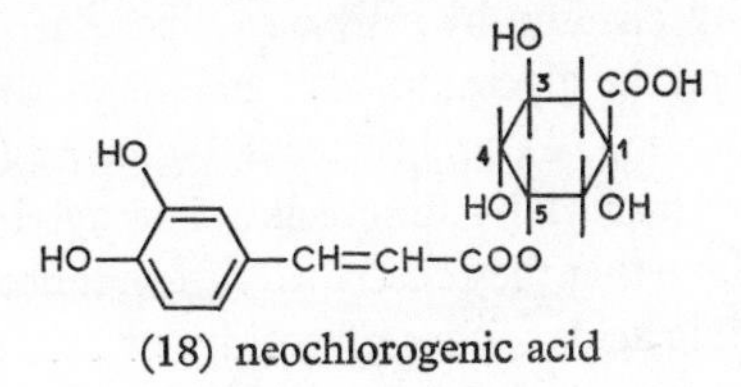

(18) neochlorogenic acid

'band 510' by Sondheimer (1958) is 4-caffeoylquinic acid (Waiss *et al.*, 1964). Finally, cynarin (19) is 1,4-dicaffeoylquinic acid. Analogous derivatives of *p*-coumaric and ferulic acids with quinic acid have been described.

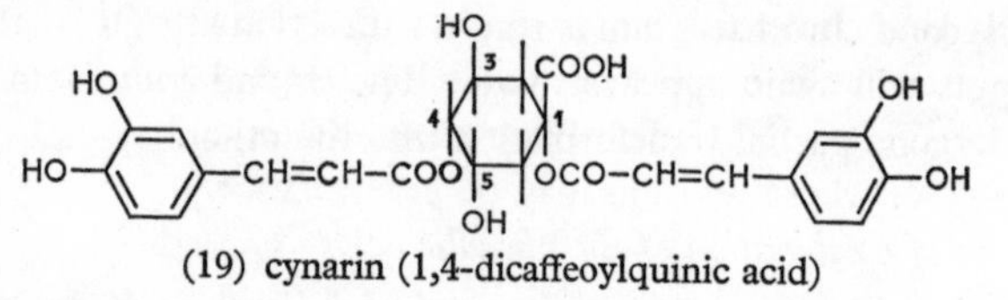

(19) cynarin (1,4-dicaffeoylquinic acid)

Acylated anthocyanins constitute another combined form of the cinnamic acids which has been known for many years. These have been studied in detail by Harborne (1964a). An example is the pigment petanin (6.2), a rhamnosyldiglucoside of petunidin to which a molecule of *p*-coumaric acid is attached by an extra linkage to a glycosidic hydroxyl. Acylated glycosides of the flavonols are less common (5.2).

By and large, it can be said that, apart from the acylated anthocyanins, the best known derivatives of the cinnamic acids are their compounds with quinic acid; but as Bate-Smith (1962a) has pointed out, it remains to be seen whether these are the commonest. In fact, recent studies have revealed the existence of new types which, since their chromatographic properties are similar to those of the quinic acid derivatives, can easily be mistaken for chlorogenic and its relatives. In particular, the importance of compounds with sugars has recently been established. Such compounds have of course been known for some time, melilotoside, a glucoside of *o*-coumaric acid, having been identified by Charaux in 1925, and an ester of *p*-coumaric acid with fructose, pajaneelin (21), by Kameswaramma and Seshadri in 1947; the latter substance is the only known phenolic compound containing a ketose (Harborne, 1964b). To appreciate the importance of these combinations, it is necessary to consult the papers by Harborne and Corner (1960, 1961). These authors have identified, for the most part in the potato, glycosides such as caffeic acid 3-glucoside (22), resembling meliltoside (20) in its structure. They also identified numerous sugar esters analogous to pajaneelin, decomposed by alkaline hydrolysis. These are considered to be more important than the glycosides and include *p*-coumaroyl-1-glucose (23), caffeoyl-1-glucose (24), and feruloyl-1-glucose (25), but caffeoyl-1-gentiobiose, *p*-coumaroyl-1-rutinose and sinapoyl-1-glucose (26) were also identified. Altogether they described fifteen different compounds of sugars with cinnamic acids, either glycosides or esters.

The results of Harborne and Corner (1961) show that the chromatographic and spectrophotometric properties of these esters of the cinnamic acids with sugars are very close to those with quinic acid. It will be shown (4.7) that the identification of the cinnamic acid moiety is relatively easy, but that the characterisation of the residue to which the acid is attached is much more difficult. It easy to understand that mistakes can be made in identifying these compounds and there are numerous cases where revision may be necessary.

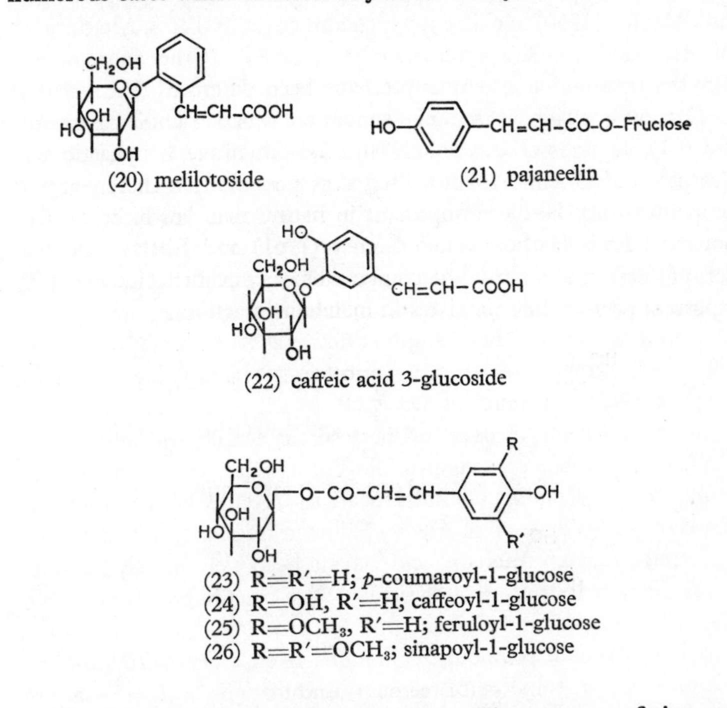

(20) melilotoside

(21) pajaneelin

(22) caffeic acid 3-glucoside

(23) R=R′=H; *p*-coumaroyl-1-glucose
(24) R=OH, R′=H; caffeoyl-1-glucose
(25) R=OCH_3, R′=H; feruloyl-1-glucose
(26) R=R′=OCH_3; sinapoyl-1-glucose

One such case is that of the grape vine. The presence of cinnamic acid derivatives in grapes has been known for many years, and it has been understood that these consist of chlorogenic acid and its isomers (Sondheimer, 1958), and compounds of quinic acid with *p*-coumaric and ferulic acids (Masquelier and Ricci, 1962). However, Weurman and Rooij (1958) found reason to suppose that the most important derivative of caffeic acid was not chlorogenic acid, and following this up, the author (Ribéreau-Gayon, 1965) has shown that the cinnamic acid derivatives of the grape are esters of *p*-coumaric, caffeic and ferulic acids with tartaric acid (27-29).

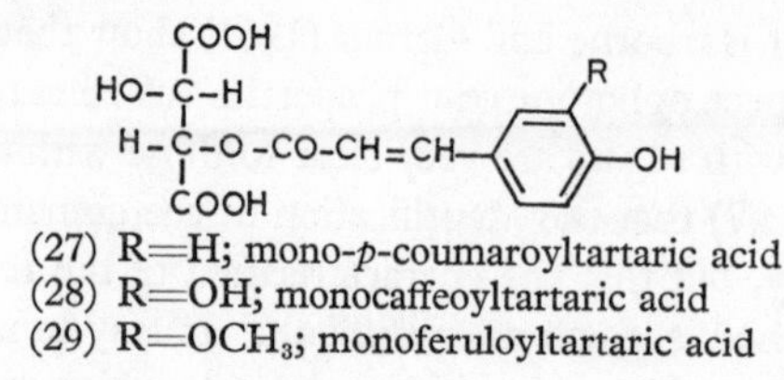

(27) R=H; mono-*p*-coumaroyltartaric acid
(28) R=OH; monocaffeoyltartaric acid
(29) R=OCH_3; monoferuloyltartaric acid

Monocaffeoyltartaric acid had been previously identified by Scarpati and d'Amico (1960) in chicory, which also contains dicaffeoyltartaric or chicoric acid (30) (Scarpati and Oriente, 1958a). Other more or less complex cinnamic acid derivatives have been described (for a review see Harborne, 1946b). One might mention as one example rosmarinic acid (31), an ester of caffeic acid and 3,4-dihydroxyphenyllactic acid (Scarpati and Oriente, 1958b). It seems possible that these complex compounds may be more important in nature than has hitherto been thought. Both Harborne and Corner (1961) and El-Basyouni and Neish (1966) consider that the compounds of cinnamic acids may play important parts as intermediates in metabolic reactions.

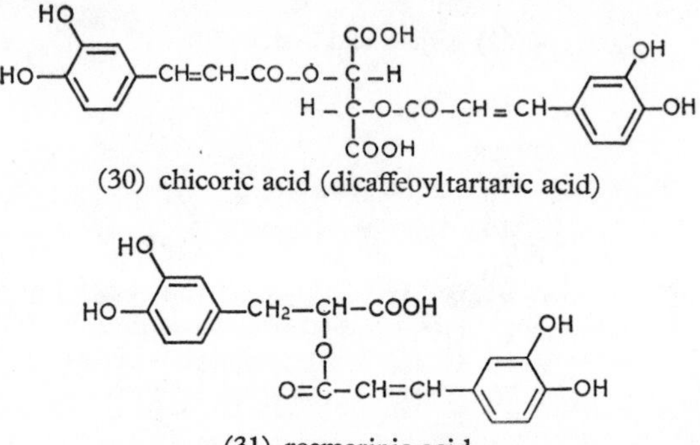

(30) chicoric acid (dicaffeoyltartaric acid)

(31) rosmarinic acid

Benzoic acids The information available about the forms of combination in which these acids occur is much less than in the case of the cinnamic acids. Apart from gallic acid, which, together with its dimer ellagic acid, is an important constituent of the hydrolysable tannins (7.2, 7.3), little is known of their combined forms. Some of them, e.g. *p*-hydroxybenzoic, protocatechuic and vanillic, occur as glycosides (Ibrahim and Towers, 1960; Bate-Smith, 1962), but it is almost certainly as acylating groups, i.e. as esters, that they occur most frequently. It is generally recognised that they enter into the composition of lignin;

at least it would be more true to say that they are liberated together with other phenolic compounds when lignin is treated with alkali.

According to Pearl (1958), lignin contains benzene nuclei engaged either in aldol (32) or dehydroaldol (33) structures, or in structures of the acyloin (α-ketol) type (34). Treatment with alkali yields an aldehyde in the case of the former (e.g. vanillin, 35)) or an acid (e.g. vanillic acid (36)) in the case of the acyloins. These acids also occur in soil humus (Jaquin, 1963) probably arising from the degradation of plant lignins.

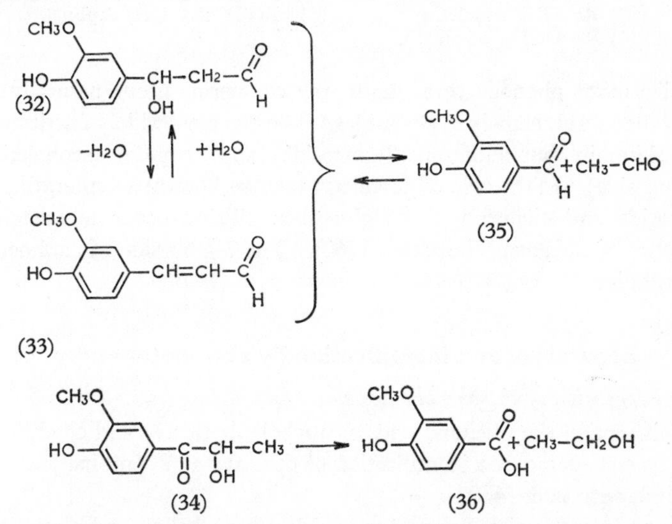

It is unlikely that the benzoic acids present in lignin account for the amounts found in plant tissues. For example, grape skins contain substances soluble in water which, when hydrolysed by alkali, yield several benzoic acids (Ribéreau-Gayon, 1964a). These acids have not, however, been reported as acylating residues in complex flavonoids. The existence of phenolic acid-protein complexes has been demonstrated by Alibert *et al.* (1968) in the leaves of *Quercus pedunculata.*

4.4. *The coumarins*

The different coumarins may be regarded as deriving from the *ortho*-hydroxycinnamic acids in the same way that coumarin itself (14) derives from *o*-coumaric acid (13). It may be recalled that coumarins, isocoumarins and chromones are structurally related (1.2, 1.8).

Many coumarins are known (Dean, 1963) but their distribution in nature is very limited. The commonest are umbelliferone (37), aesculetin (38) and scopoletin (39), the substitution patterns of which correspond, respectively, to *p*-coumaric, caffeic and ferulic acids. Fraxetin (40) and daphnetin (41) are also quite common.

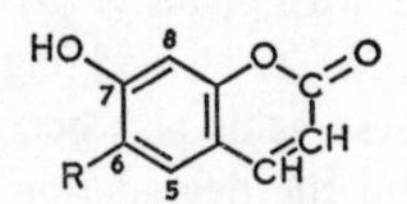

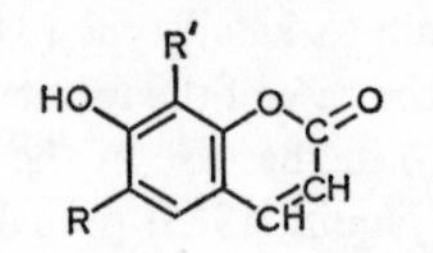

(37) R=H; umbelliferone
(38) R=OH; aesculetin
(39) R=OCH_3; scopoletin

(40) R=OCH_3, R'=H; fraxetin
(41) R=H, R'=OH; daphnetin

Like other phenolic compounds, the coumarins occur naturally in combined form, namely as glycosides. The two most widely distributed are skimmin (umbelliferone-7-glucoside) and scopolin (scopoletin-7-glucoside). In the case of aesculetin two are known: aesculin, the 6-glucoside and cichoriin, the 7-glucoside. These occur together in varieties of *Solanum* (Harborne, 1960). The 7-glucoside of daphnetin is daphnin.

Separation and identification by chromatography

4.5. *Separation of the phenolic acids*

In this section the phenolic acids as a whole, both free and combined, will be considered; the identification of their form of combination will be dealt with later (4.7).

The preparation of the extracts which, after alkaline hydrolysis and extraction with ether, are used for the separation of the acids has already been described (3.5, 3.6). When it is simply a question of examining the phenolic acids, these may be purified by extraction from the ethereal solution with 5% sodium bicarbonate, which extracts only the acids, leaving other phenols in the ether. The acids are then recovered by acidifying the bicarbonate solution and re-extracting with ether. This method of purification is not always easy, however. If the bicarbonate solution is not sufficiently alkaline, the acids will not be completely dissolved; if too alkaline, the other phenols, which are weakly acidic in character, will be partly dissolved.

The cinnamic acids have been much more studied than the benzoic acids, but it is important to consider the two classes together, because they separate on the same chromatographic paper and are revealed by the same reagents. The R_f values of the cinnamic acids in various solvents have been published, some by Bate-Smith (1954c) and others by Williams (1955). Their results are combined in Table 8. These R_f values relate to the four most useful solvents (3.6): toluene–acetic acid, Forestal, butanol–acetic acid and 2% aqueous acetic acid.

Table 8. R_f*s of cinnamic acids*
(Bate-Smith, 1954c; Williams, 1955)

Solvents	Acids			
	p-coumaric	caffeic	ferulic	sinapic
Toluene-acetic	0·05	0	0·25	0·11
Forestal	0·84	0·78	0·86	0·86
Butanol-acetic	0·88	0·78	0·84	0·79
2% acetic acid	0·40, 0·73	0·30, 0·63	0·38, 0·65	

Forestal solvent and above all toluene–acetic acid give the best separations. They also separate the benzoic acids. In the case of 2% acetic acid, Table 8 gives two values for R_f corresponding to the *cis* and *trans* isomers which are separated uniquely in this solvent (4.2). The methods of detecting the phenolic acids on chromatograms are described later on in this chapter (4.6).

Even with the use of these four solvents, however, it is not possible to separate accurately the mixture of the four commonest cinnamic acids and the seven commonest benzoic acids. There are two methods which enable this to be done. That of Ibrahim and Towers (1960) is easy to use, but although the authors separated a large number of substances present in a complex mixture, there is a danger of the separations not being adequate when the extract contains many constituents. For this reason, Ribéreau-Gayon (1964a) used another method which achieves a better resolution of the different acids, but which is more laborious to carry out.

Ibrahim and Towers' method (1960) This method consists of two-dimensional chromatography on Whatman No. 1 paper using in the first direction the supernatant phase of benzene–acetic acid–water (6 : 7 : 3) (with this solvent it is best not to equilibrate the paper with the aqueous phase since this results in trailing of the spots); and in the second direction, a solution of sodium formate (10 parts) and formic acid (1 part) in 200 of water. Although the solvent front is irregular, this does not affect the separation. The results are shown in

Fig. 7.* The spots are detected by their fluorescence and by spraying with a diazo derivative (*p*-nitroaniline or sulphanilic acid) or ferric chloride (4.6). *Cis* and *trans* isomers are separated in the second (aqueous) solvent, but not apparently, according to Ibrahim and Towers,

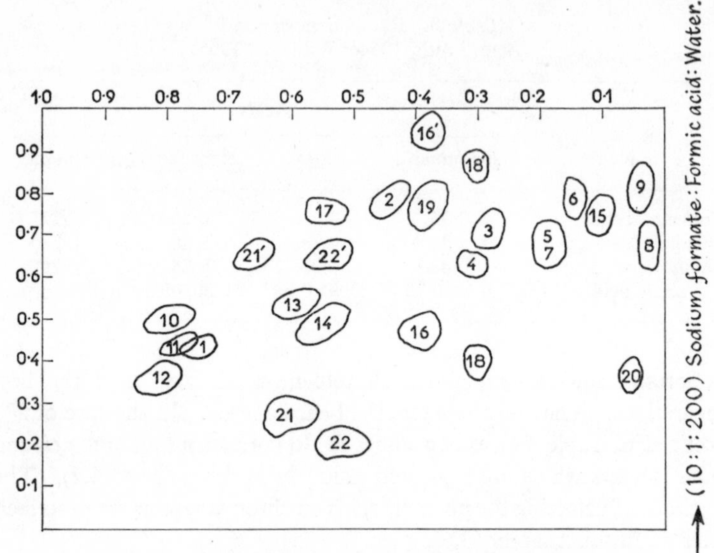

Fig. 7. *Two-dimensional chromatogram of plant phenolic acids. The acids are: 1. salicylic, 2. salicyl alcohol, 3. p-hydroxybenzoic, 4. 2, 3-dihydroxybenzoic, 5. β-resorcylic, 6. gentistic, 7. γ-resorcylic, 8. protocatechuic, 9. α-resorcylic, 10. 2-hydroxy-4-methoxybenzoic, 11. 2-hydroxy-5-methoxybenzoic, 12. 2-hydroxy-6-methoxybenzoic, 13. vanillic, 14. syringic, 15. 3-hydroxy-5-methoxybenzoic, 16. o-coumaric, 17. melilotic, 18.* p-*coumaric, 19. phloretic, 20. caffeic, 21. ferulic, 22. sinapic. (Reproduced from Ibrahim, R.K., and Towers, G. H. N.,* Arch. Biochem. and Biophys. *87, 125, 1960 by permission of the authors and Academic Press Inc.)*

in the first, based on benzene. The author himself, however, has achieved a separation of the two isomers of caffeic acid (R_f 0·05 and 0·10) after chromatographing for 72 hr.

Ribéreau-Gayon's method Using a solution of the phenolic acids in ether, three chromatograms are run as follows:

Chromatogram No. 1 The solvent is the supernatant phase of a mixture of toluene–acetic acid–water (4 : 1 : 5), as proposed by

* Separations in this system are much improved if carried out on thin layers of microcrystalline cellulose instead of on paper. (Translator's note—J.B.H.)

Bate-Smith (1954c). The time of running is 6 hr. It is essential to equilibrate the paper beforehand for 12 hr with the aqueous phase of the mixture. Salicylic, ferulic, vanillic, syringic, sinapic and *p*-coumaric acids are mobile under these conditions (Table 9). Syringic acid and sinapic acid are not however, separated in this solvent, but the ultraviolet fluorescence of the latter and their different behaviour with spray reagents enable them to be distinguished. The distinction should not, however, be attempted until after chromatogram No. 3 has been run.

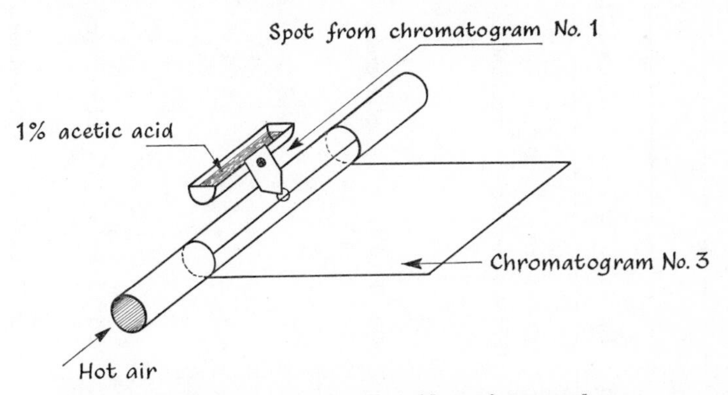

Fig. 8. *Transference of a phenolic acid spot from one chromatogram to another. 1% acetic does not elute most of the impurities present*

Chromatogram No. 2 The same solvent is used under the same conditions, but the development time is 48 hr. *p*-Coumaric, gentisic and caffeic acids are then separated (Table 9).

Chromatogram No. 3 After developing the first chromatogram but before spraying with reagents, the spot at the origin is cut out. This contains gallic, protocatechuic and occasionally gentisic acids, also various impurities including, in particular, flavonoid pigments. These constituents are transferred to another paper by the method illustrated in Fig. 8, using 1% acetic acid as solvent (Feenstra *et al.*, 1963), which has the advantage of transferring almost exclusively the phenolic acids. This procedure is, in the present case, more effective than two-dimensional chromatography.

The solvent used for chromatogram No. 3 is benzene–acetic acid–water (6 : 7 : 3) as used by Ibrahim and Towers (1960). The time of running is 72 hr. In this solvent the R_f of gentisic acid is 0·12. Caffeic acid forms two spots, as it does in aqueous solvents

Table 9. *Separation and identification of phenolic acids by chromatography* (P. Ribéreau-Gayon, 1964a)

No.	Formula	Identification	R_f	Colour			
				U.V.	U.V. + NH_3	*p*-nitroaniline	*p*-nitroaniline + Na_2CO_3
1st *chromatogram* (toluene, 6 hours)							
1	C_6H_4—COOH, OH	salicylic	0·76	blue-purple	unchanged		
2	CH_3O, HO—C_6H_3—CH=CH—COOH	ferulic	0·40	blue	blue-green	pink	pale blue
3	CH_3O, HO—C_6H_3—COOH	vanillic	0·34			yellow	violet
4	CH_3O, HO—C_6H_2—COOH, CH_3O	syringic	0·24			yellow-orange	blue
5	CH_3O, HO—C_6H_2—CH=CH—COOH, CH_3O	sinapic	0·24	blue	blue-green	pink	faint blue-green
6	HO—C_6H_4—CH=CH—COOH	*p*-coumaric	0·10	colourless	blue-violet	yellow-brown	blue

Table 9 (*contd.*)

No.	Formula	Identification	R_f	Colour			
				U.V.	U.V. + NH_3	*p*-nitroaniline	*p*-nitroaniline + Na_2CO_3
2nd *chromatogram* (toluene, 48 hours)							
7	HO—C_6H_4—COOH	*p*-hydroxy-benzoic	0·06			pale yellow	red
8	(HO)(OH)C_6H_3—COOH	gentisic	0·02	blue-green	unchanged	grey	grey-white
9	(HO)HO—C_6H_3—CH=CH—COOH	caffeic	0·003	blue	bright blue-green	yellow-brown	grey-blue
3rd *chromatogram* (benzene, 72 hours)							
10	(HO)HO—C_6H_3—COOH	protocatechuic	0·08			yellow	grey blue
11	(HO)HO—C_6H_2—COOH (HO)	gallic	0·005			yellow	beige

(Williams, 1955), with R_f 0·05 and 0·10. These two acids do not interfere with the identification of protocatechuic and gallic acids (R_f 0·08 and 0·005). In spite of these low R_f values, the running time of 72 hr results in the compounds moving a considerable distance from their origin—40 cm for gentisic acid, 28 and 16 cm for the two isomers of caffeic acid, 23 cm for protocatechuic acid and 3 cm for gallic acid.

4.6. *Identification of the phenolic acids*

This is accomplished without difficulty by comparison of R_f values and colour reactions with those of reference compounds. All these acids are available commercially, except sinapic acid which can be prepared from mustard seed by treatment with alkali and extraction with ether after acidification. Alternatively, it can be synthesised, like the other cinnamic acids, by condensation of malonic acid with the corresponding benzaldehyde (Robinson and Shinoda, 1925).

The acids are detected on the paper either by viewing in ultraviolet light or by coupling with diazonium salts, usually diazotised *p*-nitroaniline (2.8), which yields products with characteristic colours. These two procedures, together with R_f values, are sufficient for the identification of any of the naturally occurring phenolic acids, without having recourse to other reagents such as sulphanilic acid, benzidine or ferric chloride. The identification may, however, be confirmed by spectrophotometry in the ultraviolet.

Fluorescence (3.10) The four cinnamic acids and the two benzoic acids substituted in the *ortho* position (salicylic and gentisic acids) have a blue fluorescence, becoming green when fumed with ammonia, when illuminated with Wood's light (λ = 366 nm). The colours are described in Table 9.

If, instead of observing the fluorescence at room temperature, the papers are examined at low temperature, the intensities of the fluorescent colour are much increased. This procedure is easily carried out by pouring liquid nitrogen (at − 196°C) into a cardboard carton of the same dimensions as the chromatograms into which they are plunged immediately before observation. Under these conditions the benzoic acids acquire fluorescent properties which they do not possess at ordinary temperatures. These acids, but not the cinnamic acids, also acquire the property of phosphorescence in ultraviolet light of lower wavelength (253·7 nm) so that after irradiation in the ultraviolet they continue to emit a visible luminosity when irradiation is discontinued.

Reaction with diazotised p-*nitroaniline* The reagent is prepared immediately before it is required as described by Swain (1953). To 2 ml of a 0·5% solution of *p*-nitroaniline in 2N HCl are added 5 drops of 5% sodium nitrite and 8 ml of 20% sodium acetate. Spraying with this reagent gives a different colour with each acid, the colour changing when sprayed with 15% sodium carbonate. The results are given in Table 9.

Some precautions are necessary to get the best results with this reagent. First, the chromatogram should not be fumed with ammonia before the reagent is used. Second, the first spraying with the reagent should be light. Then, after spraying with Na_2CO_3, the diazo reagent can be sprayed again. Finally, fuming with ammonia after spraying improves the detection of protocatechuic and gallic acids.

To give a simple presentation of the results in quantitative terms, the intensity of the spots can be compared with those obtained when a mixture of acids of known concentration is run under the same conditions. A more exact estimate could of course be obtained by using a recording photometer as has been used (Ribéreau-Gayon, 1959) for the anthocyanins. Such a method has been described by Metche *et al.* (1962). Since, however, difficulties have been encountered in obtaining a uniform response on different chromatograms, it has been found necessary to calibrate each chromatogram, which makes the method very complicated under practical conditions. Jacquin (1963) has reported that it is necessary to spray heavily with the diazo reagent in order to obtain constant results from one chromatogram to another, but in these conditions the sensitivity is decreased.

Spectroscopy in visible and ultraviolet light The procedure for obtaining the absorption spectra of phenolic compounds after separation by paper chromatography has already been described (3.16). This can be done either by cutting out the spot and making the measurement directly, or, preferably, by eluting the spot from the paper. It is also useful to measure the absorption spectrum not only in a neutral medium (usually 95% ethanol) but also in alkali (usually 0·002 M ethanolic sodium ethylate) and in the presence of aluminium chloride.

It is important to recognise that the treatment with alkali which is necessary to hydrolyse the natural compounds of these acids may affect their absorption spectra. For instance, Geissman and Harborne (1955) observed in the case of *p*-coumaric acid kept for 18 hr in 2N caustic soda that the absorption maximum changed from 335 to 295 nm in weakly alkaline solution. The author has made similar observations.

The cinnamic acids absorb in two regions in the ultraviolet, one maximum occurring in the range 225-235 nm, and another between 290 and 330 nm. According to Harborne (1964d), the double absorption band observed in the 290 to 330 region is due to the two *cis* and

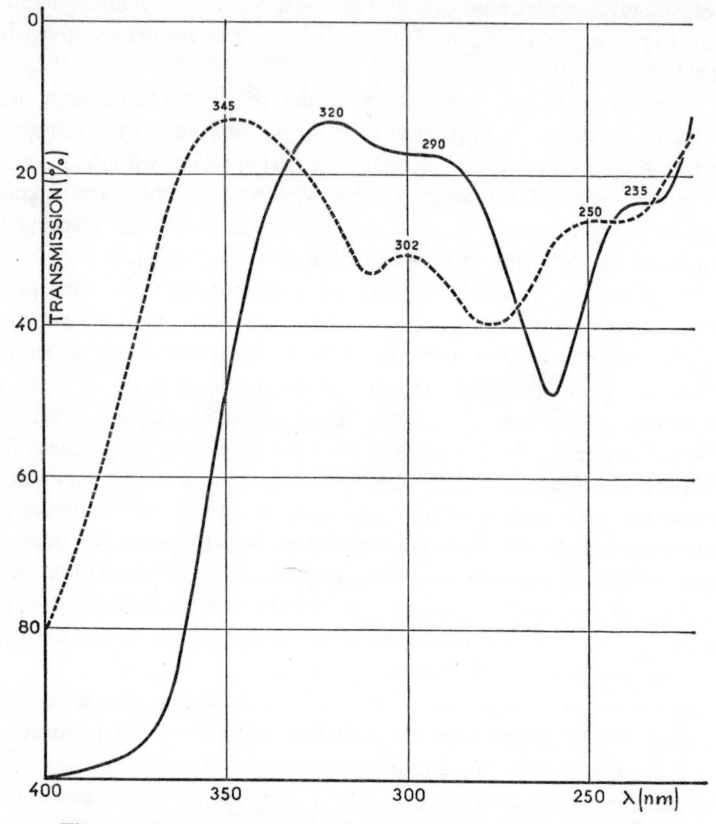

Fig. 9. *Spectral absorption of caffeic acid in 95% EtOH in 0.002M NaOEt*

trans isomers, the magnitude of one or the other depending on the preponderance of the particular isomer. These three maxima, especially those between 290 and 330 nm, are displaced towards longer wavelengths in alkaline media. As an example, the spectra of caffeic acid in neutral and alkaline media are shown in Fig. 9. The absorption maxima of the principle cinnamic acids are shown in Table 10.

In the case of benzoic acids, the absorption spectra are not necessary as an aid to identification. The values for the absorption maxima found by the author are, however, given in Table 10 by way of illustration. It can be seen that the values are profoundly affected by the hydroxylation pattern in these acids, more so than in the case of the cinnamic acids. On the other hand, methylation of hydroxyl groups has little effect on the spectral properties.

Table 10. *Absorption maxima of principal natural phenolic acids in ethanol*

Acid	λ_{max} (nm)	
p-coumaric	222	290-310
caffeic	235	290-320
ferulic	232	290-317
p-hydroxybenzoic	265	
protocatechuic	260	295
vanillic	260	290
gallic	272	
syringic	271	
salicylic	235	305
gentisic	237	335

4.7. *Examination of phenolic acid derivatives*

Since so little is known of the combined forms of the benzoic acids, consideration will be restricted to those of the cinnamic acids, and even here it is only the simplest forms of combination which will be dealt with. This is a difficult enough problem because, it must be repeated once again, different derivatives of the same cinnamic acids often have closely similar chromatographic properties. It is necessary in the first place to find a solvent which will achieve a satisfactory separation; neither butanol–acetic acid nor dilute acetic give adequate results.

It is unavoidable, therefore, to isolate each compound and to study its response to acid (i.e. glycosidic) or alkaline (i.e. ester) hydrolysis. It is then necessary to identify not only the cinnamic acid moiety but also the other partner in the combination. This is more difficult, and many authors have not done so, some, in fact, having been content merely to report the R_f of the combination.

Chromatography Harborne and Corner (1961) recommend the following four solvents:

A—butanol–acetic acid

B—water

C—*n*-butanol–2N NH_4OH (1 : 1)

D—*n*-butanol–ethanol–water (4 : 1 : 2·2)

The author (Ribéreau-Gayon, 1965) has used, together with butanol–acetic acid:

E—supernatant layer of *iso*butylmethylketone–formic acid–water (3 : 1 : 2)

F—supernatant layer of ethyl acetate–pyridine–water (2 : 1 : 2)

The solvent E was described by Jan and Reid (1959); solvent F was suggested by Forsyth and Quesnel (1957a) for the separation of sugars and applied by Loche and Chouteau (1963) to the separation of the cinnamic acid derivatives in tobacco. This solvent gives irregular R_fs dependent upon the conditions, and especially so as between ascending and descending chromatography. Using all six solvents, the different derivatives of the cinnamic acids can always be separated.

In Tables 11 and 12, the chromatographic and spectrophotometric properties of some cinnamic acid derivatives are compared with the corresponding quinic acid derivatives (e.g. chlorogenic acid) (Harborne and Corner, 1961; Ribéreau-Gayon, 1961). In addition, Harborne and Corner's work gives information about the ferulic acid and sinapic acid derivatives, not recorded in Table 12. All these substances are sugar derivatives, but their exact structure has not in every instance been determined. Their position on chromatograms is ascertained without difficulty because of their fluorescence.

To identify these compounds unequivocally it is essential to isolate them by preparative chromatography (3.14). The pure substances so obtained are then identified by their fluorescence, their absorption spectra and ultimately by their hydrolysis products.

Fluorescence and absorption spectra It can be seen in Tables 11 and 12 that the fluorescences of the esters of the cinnamic acids are broadly similar to those of the free acids (Table 9). The characteristics of the glycosides are different; thus the various esters of a particular cinnamic acid all have similar absorption spectra, irrespective of the alcoholic moiety (quinic acid, sugar, tartaric acid), but the spectra of the glycosidic derivatives are very dependent on the position of attachment of the sugar residue (Table 12).

Identification of the hydrolysis products It is first necessary to determine the response to acid and alkaline hydrolysis as previously described (3.4 and 3.5) and also enzymatically with the help of β-glucosidase (Harborne and Corner, 1961). In this way it should be possible to ascertain the nature of the chemical linkage in every instance. The identification of the cinnamic acid presents no difficulty. The solution

Table 11. *Properties of tartaric acid derivatives of cinnamic acids* (P. Ribéreau-Gayon, 1965) (Solvents A, E, F: cf. text)

Acid	Fluorescence		R_f			Absorption maxima (nm)	
	U.V.	U.V. + NH_3	A	E	F	EtOH	EtONa 0·002 *M*.
chlorogenic	blue	green	0·57	0·42	0·76	330, 300, 245	387, 312, 265
monocaffeoyltartaric	blue	green	0·55	0·53	0·44	330, 300, 245	385, 312, 265
monoferuloyltartaric	blue	blue-green	0·64	0·70	0·48		
mono-*p*-coumaroyltartaric	colourless	violet	0·69	0·74	0·50	311, 297, 225	372, 312, 230
dicaffeoyltartaric	blue	green	0·78	0·80	0·76		

Table 12. *Properties of sugar derivatives of cinnamic acids* (Harborne and Corner, 1961)
(pC: combination with *p*-coumaric; CA: combination with caffeic acid; solvents A, B, C, D: cf. text)

Designation	Nature	R_f				Fluorescence		Absorption maxima (nm)	
		A	B	C	D	U.V.	U.V. + NH_3	EtOH	EtONa 0·002 *M*
p-coumaroylquinic	ester	0·71	0·92	0·09	0·54	colourless	blue		
pC 1	ester	0·66	0·73–0·81	0·30	0·75	colourless	blue	235, —, 313	240, 365
pC 2	ester	0·49	0·84	0·16	0·64	colourless	blue	232, —, 314	240, 364
caffeoylquinic (chlorogenic acid)	ester	0·59	0·67–0·84	0·01	0·50	blue	green		
CA 1	ester	0·51	0·61–0·72	0·08	0·63	blue	green	250, 305, 334	265, 400
CA 2	glycoside	0·46	0·46	0·02	0·55	blue	intense purple	228, 295, 311	235, 335
CA 3	glycoside	0·29	0·77–0·83	0·00	0·46	blue	turquoise	—, 300, 322	247, 365
CA 4	ester	0·29	0·77–0·83	0·03	0·46	blue	green	245, 305, 335	265, 400

after hydrolysis is extracted with ether (after reacidification in the case of alkaline hydrolysis) and the ethereal extract is chromatographed as described previously (4.5).

In order to identify the other residue participating in the combination, it is first necessary to remove excess acid or alkali. In the case of the former it is a question of glycosidic combination, and the method used is that described for the glycosides of flavones and anthocyanins (5.9, 6.15). In the second case, excess of soda is removed by the use of a cation exchange resin. After concentrating, the hydrolysis product (sugar, quinic acid, tartaric acid, etc.) is identified by using the classical techniques of chromatography (Lederer and Lederer, 1957; Lederer, 1959, 1960).

The author's own procedure is as follows: the eluate from the paper chromatogram is evaporated to dryness *in vacuo*. 2 ml of 2N NaOH are added in a current of nitrogen, and the stoppered container kept for 4 hr at room temperature. 4 ml of Dowex 50 H^+ form (i.e. approximately twice the theoretical amount to neutralise the NaOH) are added, the mixture shaken for a few minutes, and the free fluid decanted and filtered. The resin is washed with 2 ml of water, which are added to the filtrate. The filtrate is concentrated *in vacuo*, then chromatographed by the usual methods, so as to identify the substances present. Needless to say, the freedom of the eluate from impurities must be assured by chromatography in a number of solvents before the hydrolytic procedure is carried out.

Note: Sondheimer (1958) has described a method of separating caffeic acid and its various compounds with quinic acid (chlorogenic acid and its isomers) by column chromatography on silicic acid. Using this method, Sondheimer found, in 100 g of Steuben grapes, 140 mg of chlorogenic acid, 20 mg of isochlorogenic acid, 2 mg of neochlorogenic acid and 20 mg of an unidentified isomer ('Band 510'). These results do not agree with recent results of Ribéreau-Gayon (1965). It is possible that column chromatography does not differentiate between the esters of quinic and tartaric acids. There may also be other types of combination of the cinnamic acids.

4.8. *Examination of the coumarins and their glycosides*

Chromatographic studies of the coumarins are much less numerous than those of the cinnamic acids. The most detailed work is that of Swain (1953) and Harborne (1960b). Bate-Smith (1954c) has shown that umbelliferone and scopoletin, i.e. the coumarins corresponding to

Table 13. *Properties of common coumarins and their glycosides*
(Swain, 1953; Bate-Smith, 1954c; Harborne, 1960b)

Designation	R_f						Fluorescence	
	toluene-acetic	butanol-acetic	acetic acid 10%	acetic acid 15%	water	butanol-NH_4OH 2*N* (1 : 1 v/v)	U.V.	U.V. $+NH_3$
coumarin	0	0·92	0·76		0·67			
umbelliferone	0·07	0·89	0·60		0·57		bright blue	bright blue
skimmin	0	0·52	0·82		0·72		pale violet	pale violet
aesculetin	0	0·79	0·45	0·56	0·28	0·13	blue	clear blue
aesculin	0	0·53	0·69	0·76	0·56	0·13	clear blue	bright blue
cichoriin	0	0·53		0·78	0·61	0·10	pale pink	orange-yellow
scopoletin	0·15	0·83	0·51	0·59	0·29	0·35	blue-violet	bright blue
scopolin	0	0·53		0·85	0·64	0·44	blue-violet	blue-violet
daphnetin	0	0·81	0·54		0·61		pale yellow	bright yellow

p-coumaric and ferulic acids, are the only ones of the commoner coumarins to migrate in toluene–acetic acid.

The principle R_f values to be found in the literature are collected together in Table 13. For any one solvent, the values may derive from two different sources. The ultraviolet fluorescent colours of the substances are also recorded. The coumarins occur in nature as glycosides (3.4) and their identification is achieved by the usual methods for phenolic glycosides (Chapter 3). So far as possible, the R_f values and colour reactions are used, complete determination being achieved by acid hydrolysis and identification of the hydrolysis products. Ultraviolet spectroscopy can also be used in the course of identification. The coumarins possess two main peaks in the ranges 310-350 nm and 220-230 nm and, in addition, one or two secondary maxima which may be reduced to shoulders (Goodwin and Pollock, 1954). The paper by Harborne (1960b) gives data on the spectra of aesculetin and scopoletin and their derivatives.

5

Flavones, Flavonols and Related Compounds

Introduction

This chapter deals with the naturally occurring polyphenols possessing the general structure $C_6—C_3—C_6$, other than the anthocyanins (1.1, 1.3, 1.4). These are the very numerous substances sometimes known as anthoxanthins. The most important of these, i.e. those which are the most widely distributed, the flavones and flavonols, will be fully discussed; less important ones, such as the flavanones, isoflavones, chalcones and aurones will be considered in less detail. The flavans (flavan-3-ols or catechins and flavan-3,4-diols or leucoanthocyanidins) will not be dealt with in this chapter: they differ in that they do not occur naturally as glycosides, but as condensed products which constitute the tannins (Chapter 7).

In the first section, the more important of the substances are described, first the aglycones and then their glycosides as well as some exceptional compounds of particular interest. This section owes much to the detailed treatments of Harborne and Simmonds (1964) and Harborne (1964b). The next section deals with the chromatographic techniques used in their examination. Then, the methods of identification of the aglycones and their glycosides are discussed, more especially the glycosides because of their greater complexity. The importance of spectrophotometry for these identifications, in the visible, the ultraviolet and the infrared is recognised by a separate section. Nuclear magnetic resonance has also been used in this connection (Massicot *et al.*, 1962, 1963; Mabry *et al.*, 1965; Hillis and Horn, 1965; Venkatamaran, 1966; Grouiller, 1966); this method can be expected to be developed further in the future.

It might seem logical, in studying plant flavanoids, to isolate each glycoside and subject it to a rigorous identification of the components, especially of the aglycone moiety. However, it is preferable in actual practice first to find out what aglycones are present, by subjecting the total plant extract to acid hydrolysis (3.4, 3.6).

The principal substances

5.1. *The aglycones*

Only the substances of commonest occurrence will be mentioned here. Further details can be found in Geissman and Hinreiner (1952), Gripenberg (1962) and Harborne and Simmonds (1964).

The flavones, apigenin (1) and luteolin (2) are widely distributed in the angiosperms. Tricin (3), on the contrary, is common only in

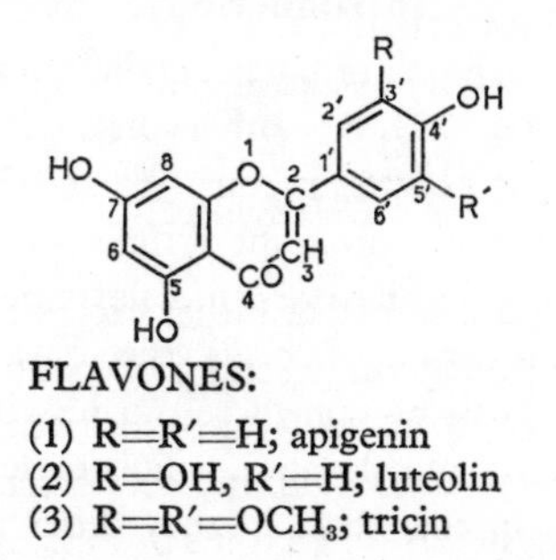

FLAVONES:
(1) R=R′=H; apigenin
(2) R=OH, R′=H; luteolin
(3) R=R′=OCH_3; tricin

grasses, which is surprising in view of the frequency in plants of other phenolic compounds having the same substitution pattern as that present in the B ring in tricin, e.g. sinapic acid and malvidin (an anthocyanidin). Just as in the case of the cinnamic acids, the flavone with the *vic*-trihydroxy substitution pattern is unknown.* Among the less common members of this class, flavone itself, which has no substituent in either benzene ring and which occurs in *Primula*, and chrysin, 5,7-dihydroxyflavone, a constituent of the heartwood of species of *Pinus*, may be mentioned. Other structures are occasionally found (Thomas and Mabry, 1968; Markham and Mabry, 1968a), for instance *bis*-flavones resulting from the condensation of the two molecules of apigenin or an apigenin derivative (Sarvada, 1958; Miura *et al.*, 1968).

The flavonols 3-hydroxyflavones are distinguished from the flavones by the presence of an OH group in position 3; this is the only hydroxyl

* This is no longer true, since 5,7,3′,4′,5′-pentahydroxyflavone (tricetin) has now been found in *Lathyrus pratensis* and in one or two gymnosperms (Malcher and Lamer, 1967). (Translator's note—E. C. B.-S.)

group in the molecule which is not phenolic. These are the most widespread of all flavonoids, quercetin (5) especially being one of the principal phenolic constituents of plants. Kaempferol (4) and myricetin (7) are also widely distributed. Methylation of the OH groups also occurs; the best known *O*-methylated derivative is isorhamnetin (6), which is present in the pollen of some plants.

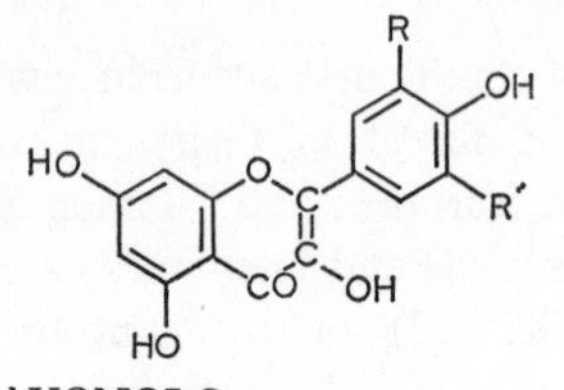

FLAVONOLS:

(4) R=R′=H; kaempferol
(5) R=OH, R′=H; quercetin
(6) R=OCH_3, R′=H; isorhamnetin
(7) R=R′=OH; myricetin

On the other hand, a number of rather exceptional structures occurs in the flavonols. Quercetagetin (8) and gossypetin (9) are characterised by an additional OH group in the 6- or 8-position. They are especially found in the Compositae and Leguminosae. Fisetin (10) and robinetin (11) lack an OH group in position 5, thus possessing a resorcinol residue in place of the usual phloroglucinol residue. They are rather uncommon, but are also present in the Leguminosae. Morin (3,5,7,2′,4′-pentahydroxyflavone) with the unusual 2′-hydroxy substitution has been found in several plants in the Moraceae.

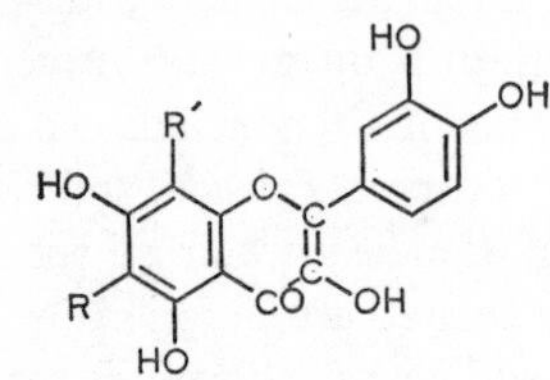

FLAVONOLS:

(8) R=OH, R′=H; quercetagetin
(9) R=H, R′=OH; gossypetin

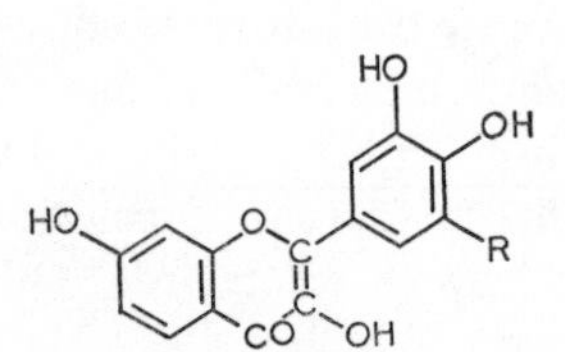

FLAVONOLS:

(10) R=H; fisetin
(11) R=OH; robinetin

The flavanones are derived from the flavones by elimination of the double bond in the central heterocycle. Naringenin (12) and eriodictyol (13), related respectively to apigenin and luteolin, are the best known. The flavanones are fairly widely distributed, but not so widely as the flavones and flavonols. The 3-hydroxyflavanones (known either as flavanonols or 2,3-dihydroflavonols), for instance taxifolin or dihydroquercetin, play an important part in schemes of flavonoid biosynthesis (8.5) (Pachéco, 1966).

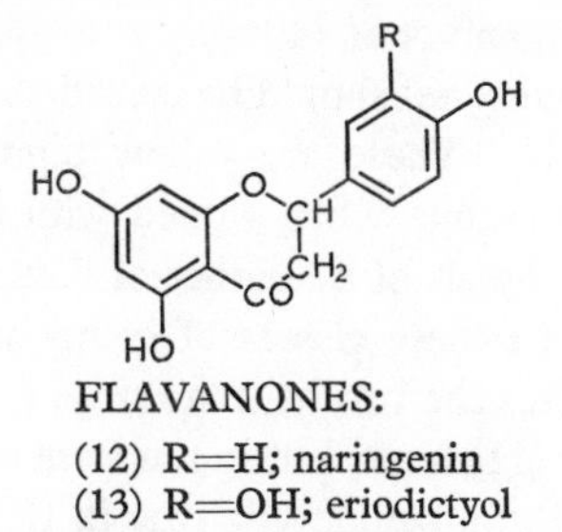

FLAVANONES:
(12) R=H; naringenin
(13) R=OH; eriodictyol

It should be noted that the flavanones have an asymmetric carbon atom at position 2 which determines that each substance is capable of existing in two optical isomeric forms. According to Horowitz (1964), hesperetin, present as the glycoside hesperidin in *Citrus*, is laevorotatory. The absolute configuration of hesperidin (14) has been demonstrated by ozonolysis, which yields L-malic acid (15) (Arthur *et al.*, 1956; Arakawa and Nakazaki, 1960).

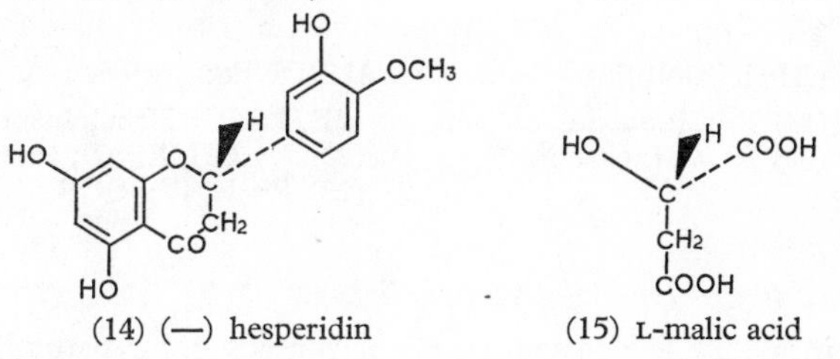

(14) (—) hesperidin (15) L-malic acid

The isoflavones are isomers of the flavones; the lateral benzene ring is attached to the carbon atom in position 3 instead of position 2. A great many different members of this class have been described (for example, daidzein (16), genistein (17) and orobol (18)), but they are

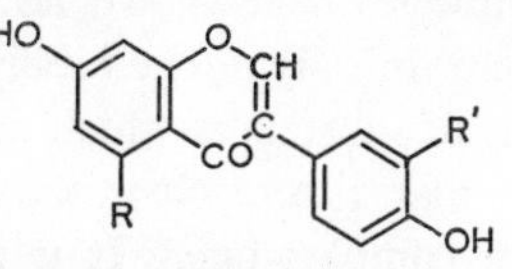

ISOFLAVONES:
(16) R=R'=H; daidzein
(17) R=OH, R'=H; genistein
(18) R=R'=OH; orobol

restricted almost entirely to the Lotoideae (Papilionatae), a subfamily of the Leguminosae (Harborne, 1967). An interesting series of isoflavones have been identified, for example, in the genus *Baptisia* (Markham *et al.*, 1968). 5-Methylgenistein has been identified in the wood of *Cytisus* by Chopin *et al.* (1964).

The chalcones and aurones should be considered together (Harborne and Simmonds, 1964). These are yellow pigments, the colours of which change to red-orange when fumed with ammonia. They are found together in the petals of a number of flowers in the Compositae (Harborne, 1966a). In these classes of compound, the three carbon atom chain which unites the two benzene rings is not in the form of an oxygen heterocycle of six atoms, but in the form of a linear chain in the case of chalcones and of a pentacyclic ring in that of the aurones. In each case only a limited number of substances are known. Butein (19) and okanin (20) are examples of chalcones and sulphuretin (21), aureusidin (22) and leptosidin (23) of aurones.

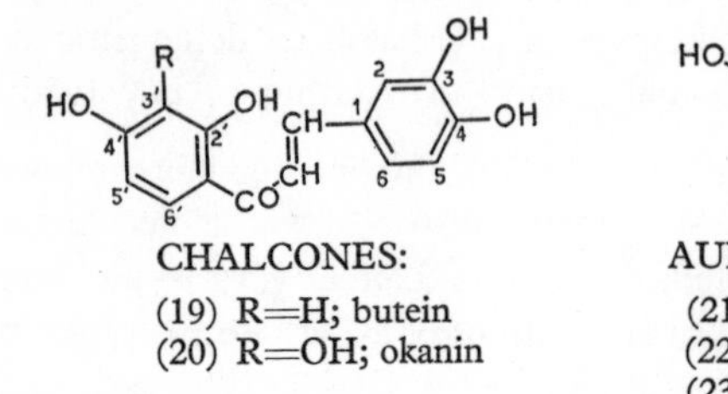

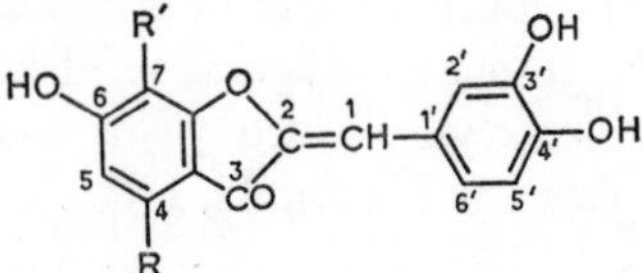

CHALCONES:
(19) R=H; butein
(20) R=OH; okanin

AURONES:
(21) R=R′=H; sulphuretin
(22) R=OH, R′=H; aureusidin
(23) R=H, R′=OCH_3; leptosidin

5.2. *The different types of glycosidic linkages*

Except in a very few instances, the flavonoids occur naturally in the form of glycosides. Before the advent of chromatography, the presence of aglycones in plant tissue was frequently reported, but these reports were almost certainly due to partial hydrolysis during the manipulation involved in the purification and identification of the substances actually present. Modern techniques have shown how easily such hydrolysis may take place, either enzymically or chemically, leading to the appearance on chromatograms of substances which do not in fact exist in the plant. This hydrolysis may also result in the appearance of a simpler glycoside from a more complex one. It is necessary to bear these possibilities in mind when carrying out an identification.

The sugars involved in the structure of phenolic glycosides are almost exclusively aldoses. The presence of D-fructose, a ketose, is quite exceptional. D-glucose is certainly the commonest sugar present; amongst others which have been identified are D-galactose, D-xylose, L-rhamnose, L-arabinose and a derivative of glucose, D-glucuronic acid.

The cyclic form of sugars is most usually the pyran form with six atoms. Combination with the aglycone is the β-configuration between C 1 of the sugar and a hydroxyl oxygen (*O*-glycoside). Formula (24), for instance, represents the 7-β-D-glucoside of apigenin. The α-configuration has only been reported in the case of arabinose, for example, the α-L-arabinoside of quercetin (25) which has the additional peculiarity of being in the furanose form.

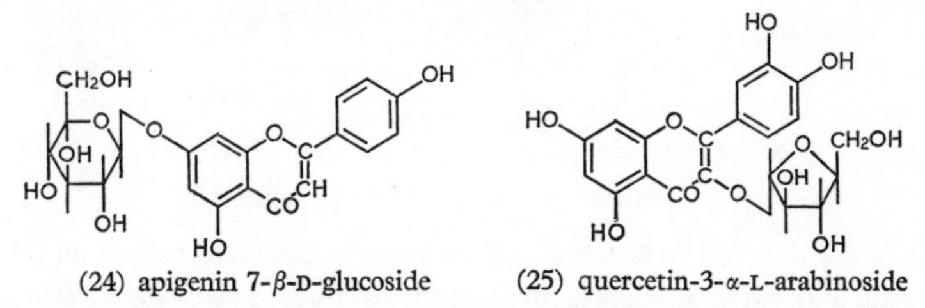

(24) apigenin 7-β-D-glucoside (25) quercetin-3-α-L-arabinoside

Besides the simple sugars so far mentioned, disaccharides and even trisaccharides may be involved in the glycosidic structures. They have been discussed in detail by Harborne (1964b), and only the more important will be described here. Sophorose consists of two molecules of glucose (Gluc β1→2 Gluc), sambubiose of one molecule of xylose and one of glucose (Xyl β1→2 Gluc) and rutinose (28) of one molecule of glucose and one of rhamnose (Rham α1→6 Gluc). The fact that rhamnose occurs in rutinose with the α-configuration has been recently established (cf. Horowitz, 1964); in many recent publications it is still given with the correct β-structure. Rutin (26), i.e. quercetin 3-rutinoside, is very widely distributed in the plant kingdom. It is extracted especially from *Quercus tinctoria* and represents an important source of rhamnose.

The different glycosidic combinations entered into by the flavonols, especially kaempferol and quercetin, are exceedingly numerous. At

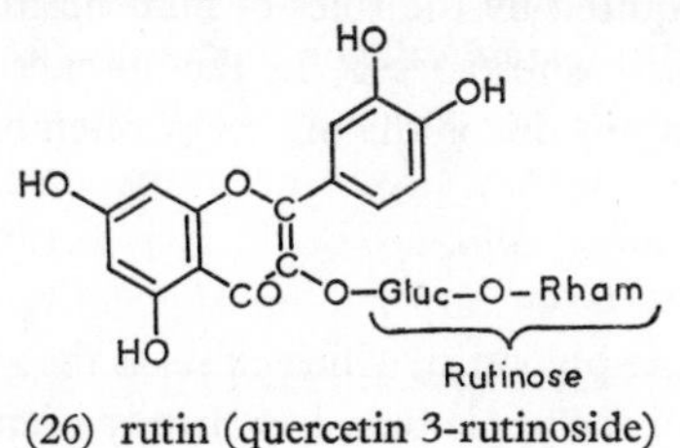

(26) rutin (quercetin 3-rutinoside)

least fifty different glycosides have been reported by Harborne (1964b, 1965a). Usually the sugar residue is attached at position 3, a second sugar residue if present, being frequently in position 7. Unlike what is so usual in the anthocyanins, glycosylation at position 5 is extremely rare, having in fact only been observed on a few occasions (Harborne, 1970).

Acylated flavonol glycosides are also known but they are much less common than in the anthocyanins of the same type (6.2). One such compound is tiliroside (27), discovered by Horhämmer *et al.* (1959).

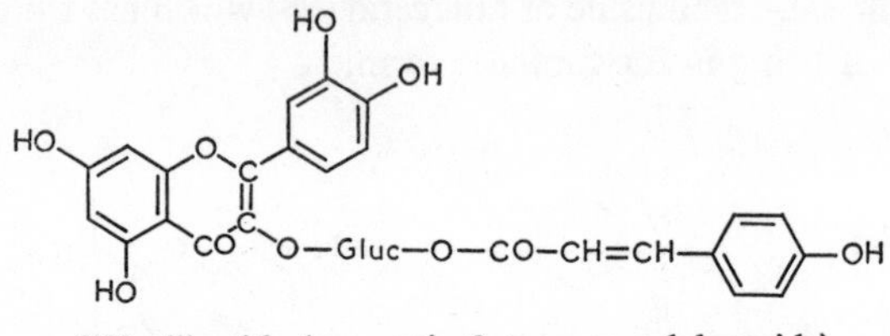

(27) tiliroside (quercetin 3-*p*-coumarylglucoside)

In the case of the flavones and flavanones, in which there is no OH group at position 3, the principal known glycosides have a sugar residue at position 7 (e.g. apigenin 7-glucoside (24)), but a number of 5-glucosides are known. Hesperidin (28) is one of the flavanone glycosides whose structure is best known, i.e. the 7-β-rutinoside of hesperitin.

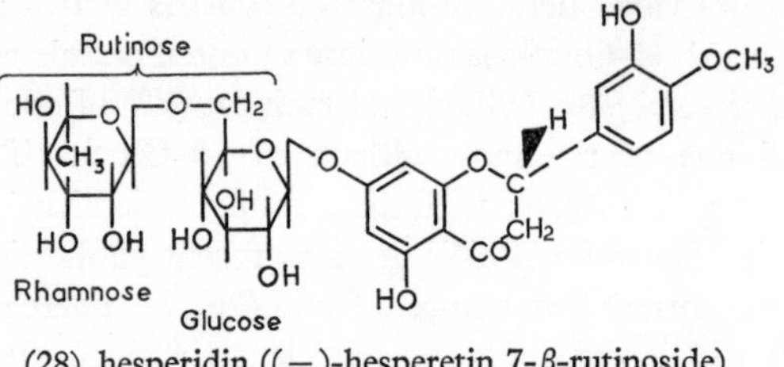

(28) hesperidin ((−)-hesperetin 7-β-rutinoside)

Isoflavones can exist in the free state, for instance in the heartwoods of some trees; the sugar residue, if present, is usually at position 7. Also in the case of the chalcones, the sugar group is in the corresponding position, but as required by the rules of nomenclature, this position is numbered 4′ in these substances. In the aurones the corresponding position is 6, and again this is the one most often occupied by a sugar residue.

5.3. *Exceptional structures*

The dihydrochalcone phloretin, which exists as the glucoside phloridzin in various tissues of the apple tree, has already been mentioned (1.4).

The most interesting of the unusual flavone derivatives is undoubtedly the *C*-glycoside, in which a sugar residue is combined, through its carbon atom 1, with an aglycone, usually a flavone, through one of the latter's carbon atoms forming a C—C bond. These substances, which have been studied especially by Horhämmer and Wagner (1961), Chopin (1966) and Bhatia and Seshadri (1967) are highly resistant to hydrolysis by acids and by enzymes.

The best known *C*-glycosides are vitexin (29), 8-*C*-glucosylapigenin, and orientin (30), 8-*C*-glucosylluteolin. Their structure should be

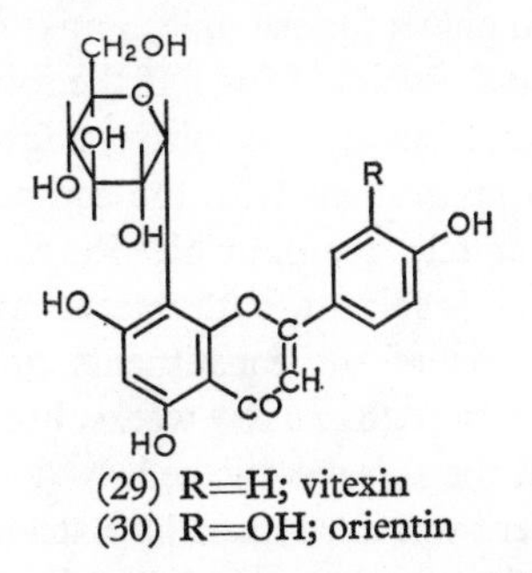

(29) R=H; vitexin
(30) R=OH; orientin

compared with that of an *O*-glucoside (24). The existence of *C*-glycosides of anthocyanins has not been definitely demonstrated (Harborne, 1964b), and there is only one known instance of a *C*-glycosylflavonol, keyakinin; this is 6-glucosyl-7-*O*-methylkaempferol (Funaoka and Tamaka, 1957; Hillis and Horn, 1966). Bergenin is a *C*-glucoside related to the isocoumarins and mangiferin the *C*-glucoside of a xanthone (Bate-Smith, 1965). The *C*-glycosides are often found combined as *O*-glycosides, e.g. the 4′-rhamnoside of vitexin.

Chromatographic techniques

In this section the application of chromatography to the flavones and related compounds is discussed. A more general discussion of this technique has been given in Chapter 2.

5.4. *Solvents*

The solvents to be described are used in pairs: butanol-acetic and Forestal for the aglycones, and butanol-acetic and aqueous acetic acid for the glycosides. In each case, pairs of solvents may be used for two-dimensional chromatography.

Butanol–acetic This is without doubt the most important solvent for the study of phenolic compounds (3.6, 3.8). In the case of the flavones it is equally useful for the aglycones and the glycosides. It was first introduced by Partridge (1947) for the separation of sugars, then by Bate-Smith in 1948 for the flavonoids. Originally the solvent was prepared by mixing 4 vols. of *n*-butanol, 1 vol. of acetic acid and 5 vols. of water. After equilibration, the supernatant layer constitutes the solvent proper. In order to simplify the technique, it is becoming more and more customary to use mixtures of the three components which provide a single phase, for example, a mixture in the proportions 6 : 1 : 2 (Nordström and Swain, 1953). If the solvent becomes cloudy owing to the separation of an aqueous phase, a few drops of acetic acid are added. Rigorous precautions with the composition of the solvent are necessary if accurately reproducible R_f values are required, a difficult problem with this solvent, but unnecessary when it is merely a question of identification of the constituents present. This solvent should not be used for more than a few weeks, because it slowly undergoes esterification and the R_f's are affected.

Forestal This solvent was introduced by Bate-Smith (1954b) for the separation of the anthocyanidins. It has subsequently proved very useful for the study of other flavonoid aglycones. It is essentially an aqueous solvent, prepared by mixing acetic acid, water and conc. HCl (30 : 10 : 3 by vol.). It owes its name to the fact that its composition was suggested to Dr E. C. Bate-Smith by the Forestal Laboratories at Harpenden. With this solvent the glycosides, and other constituents which might interfere, have high R_f's. Excellent separation is achieved for the aglycones with R_f's lower than 0·75. At higher R_f's the chromatogram is overcrowded, so that this is not a good solvent for the cinnamic acids.

Dilute acetic acid In this and similar solvents the flavonoid aglycones do not as a rule move from the start line, only those in which the central heterocycle is saturated (e.g. the flavans and flavanones) and in consequence have a non-planar configuration. In such cases, it is possible to separate optical isomers, and a particular case of this, i.e. the catechins (flavans) is described later (7.7).

On the other hand, aqueous solvents, water itself for example, enable good separations to be achieved of flavonoid glycosides (Roberts *et al.*, 1956). The addition of acetic acid has the object of avoiding trailing. Concentrations between 2 and 15% are employed. R_f values are higher, the higher the concentration of acetic acid, but the separations are unaffected. For particular problems of separation many other solvents have from time to time been proposed. Some of these are described by Harborne (1959b) and Seikel (1962).

5.5. *Examination of the chromatograms*

The general principles governing the detection of flavones and their derivatives have been described earlier (3.10, Table 5). Many of these substances are colourless but become yellow when fumed with ammonia or sprayed with alkali. This reaction is, however, not very sensitive.

Examination in ultraviolet light of wavelength 360 nm is the procedure most generally adopted, and is usually adequate for routine purposes. Apart from flavanones and isoflavones, practically all the substances in this class appear as coloured spots in ultraviolet light, some of them being fluorescent. The acid character of phenols (2.2) which involves a change in structure and colour in alkali is also used. The simplest procedure is to fume the paper in ammonia vapour when under the ultraviolet lamp, this having the advantage that the original appearance is restored when the ammonia is removed in a current of air, and the spot can be eluted for further examination. This treatment should, however, be used with caution if it is desired to isolate the component unchanged since some oxidation is likely to take place under these conditions. The enhanced luminescence at low temperatures may also be made use of by examining the chromatogram in liquid nitrogen (4.6).

The results of examination in ultraviolet light have been described in detail by Harborne (1959b) and may be summarised as follows:

(*a*) Flavonols and their glycosides in which position 3 is free appear yellow, with minor differences which are nevertheless worth taking into account (kaempferol yellow-green, isorhamnetin and quercetin yellow, myricetin golden yellow). This colour is affected very little by exposure to ammonia.

(*b*) Flavones, which do not have a hydroxyl group in position 3, and the 3-glycosides of flavonols appear yellow-brown in ultraviolet light. The colour changes to bright yellow in ammonia. Glycosides may therefore have a quite different appearance from their parent aglycones.

(*c*) Flavonols which do not have an OH group in position 5 (fisetin (10), robinetin (11)), have intense yellow or yellow-green fluorescence.

(*d*) A number of substances behave in a way not predictable from their structure. For example, quercetagetin (8), although hydroxylated in position 3, does not appear yellow in the ultraviolet but forms a dark brown spot. Quercetagetin and its isomer gossypetin are both visibly yellow on the paper.

(*e*) The isoflavones, and also the flavanones, are difficult to detect by these methods. The more important flavanones (naringenin (12), eriodictyol (13) and hesperitin (14)) do, however, appear pale yellow in the ultraviolet when fumed with ammonia.

(*f*) The aurones, which are naturally bright yellow in visible light, have the same appearance in ultraviolet light, changing to bright red in ammonia. The chalcones also are yellow in visible light, but appear brown in ultraviolet changing to red-brown with ammonia.

If examination in ultraviolet light is not conclusive, the paper may be sprayed with suitable reagents. The principal reagents used have already been described (3.10). The mixture $FeCl_3$—$K_3Fe(CN)_6$ is sensitive, but lacks specificity, as also does ammoniacal silver nitrate (2.6). With aqueous sodium carbonate colours are produced which range from yellow (flavones and flavonols) to green (isoflavones and flavanones). Myricetin, however, changes to blue, and the aurones and chalcones to red. With aluminium chloride, all the substances dealt with in this chapter react (2.4) with the formation of complexes which are coloured in ultraviolet light and sometimes also in visible light. The fluorescences of these complexes are yellow in the case of flavonols and isoflavones, green with flavones, flavanones and aurones, and orange with chalcones (Harborne, 1959b).

5.6. *Thin layer chromatography*

As indicated previously (3.3, 3.11), column chromatography, while useful as a preliminary treatment for complex mixtures, is not so efficient as paper chromatography for the isolation of individual constituents.

Thin layer chromatography has been used for the separation of the whole range of phenolic constituents by Nybom (3.11) and also applied to the flavonoids by Paris (1961, 1963), and by Stahl and Schorn (1961). Egger (1961, 1964) and Egger and Keil (1965) have described a useful method enabling the glycosidic linkages of flavonols to be identified.

In the study of phenolic compounds, thin layer chromatography must be regarded as a tool to be used in solving particular problems to provide further evidence in support of results obtained by other methods. It is unlikely that it will replace paper chromatography completely.

Paris (1961, 1963) used a layer of silicic acid prepared by spreading on a glass plate a homogeneous mixture of 25 g of silicic acid Mallinckrodt, 2 g of corn starch (to give the layer the necessary rigidity) and 75 ml of water. The solvents used are the mixtures isopentanol–hexane–acetic acid–water (3 : 1 : 3 : 3), toluene–ethyl acetate–methanol (8 : 6 : 1) and toluene–ethyl acetate–ethanol (4 : 2 : 2). Separations take 30 to 150 min. In a later publication (1968), Paris suggests the following solvents: benzene–ethyl formate–formic acid (75 : 24 : 1) and ethyl acetate–methanol–water (100 : 16·5 : 13·5).

The use of this inorganic supporting material has enabled Paris to devise a novel method of visualising the spots on the plate. Flavones and their derivatives are reduced by nascent hydrogen produced by the action of concentrated hydrochloric acid on magnesium. Coloured products are formed; for instance quercetin is reduced to cyanidin (6.5). The magnesium is mixed with the silicic acid in the form of a fine powder to the extent of 20% just before the plates are poured. The sample to be analysed is applied in a water-free medium, for instance absolute methanol. The solvent should be weakly acid, i.e. boric acid–methanol–ethyl acetate (5 : 10 : 90). After development (45 min), the plate is fumed with conc. HCl. Flavanones give violet, flavonols red and flavones orange colours.

Egger (1961) used a layer of polyamide, prepared by spreading on glass plates a mixture of 5 g of polyamide powder and 45 ml of methanol. The solvent used is a mixture of water–ethanol–acetylacetone (4 : 2 : 1), separations taking 2-3 hr. The spots are detected in ultraviolet light.

Methods of identification

5.7. *Comparison of R_f values*

In Tables 14, 15 and 16 are assembled the R_f's given in the literature for the principle flavones, flavonols, flavanones, isoflavones, chalcones and aurones, and also those of their commonest glycosides. For the chromatography of *C*-glycosides the paper by Seikel *et al.* (1966) should be consulted.

It is not possible to compare R_f's obtained experimentally directly with those reported in the literature, because they are not sufficiently reproducible (3.13). For instance, the author has found R_f's of 0·88, 0·78 and 0·54 respectively for kaempferol, quercetin and myricetin in butanol–acetic, very different from those reported in Table 14. It is essential, therefore, to run reference compounds on each chromatogram.

When these are not obtainable commercially they can be synthesised, especially the flavones (Gripenberg, 1962) and the flavonols (Robinson and Allen, 1924; Pachéco and Grouiller, 1965, 1966). However, even if reference compounds are not available, much can be learned from R_f's (3.13).

For more certain identification it is necessary to isolate the individual pigments by the methods previously described (3.14) using preparative paper chromatography. The procedures described in the following paragraphs, and the measurements of absorption spectra, are carried out on pigments purified in this way.

Table 14. *R_f's of flavone and flavonol aglycones* (Harborne, 1959b, 1965c)

	Butanol-acetic	Forestal
FLAVONES		
apigenin	0·89	0·83
luteolin	0·78	0·66
tricin	0·73	0·72
vitexin	0·43	0·82
FLAVONOLS		
kaempferol	0·83	0·55
quercetin	0·64	0·41
myricetin	0·43	0·28
isorhamnetin	0·74	0·53
fisetin	0·73	0·58
robinetin	0·40	0·36
quercetagetin	0·31	0·26
gossypetin	0·31	0·26

5.8. *Identification of aglycones*

The aglycone is identified after rupturing the glycosidic linkage between it and one or more sugar residues. Hydrolysis is carried out in 2N HCl in the water bath at 100°. The resistance to hydrolysis of different glycosidic bonds, however, varies. Harborne (1965a) has shown that hydrolysis may be complete in anything between 2 to 250 min and a kinetic study of the rate of hydrolysis may be used to identify the glycoside (5.10). Moreover, if insufficient time is allowed for hydrolysis, a

Table 15. R_f's of *flavone and flavonol glycosides* (Harborne, 1959b, 1965a, b)

	Butanol-acetic	Water	Acetic acid 15%
FLAVONES			
Apigenin			
7-monoglucoside	0·65	0·04	0·25
7-rhamnoglucoside	0·58	0·09	0·46
7-monoglucuronide	0·57	0·13	0·29
Luteolin			
7-monoglucoside	0·44	0·01	0·15
7-diglucoside	0·40	0·05	0·29
5-monoglucoside	0·82	0·00	0·07
FLAVONOLS			
Kaempferol			
3-monoglucoside	0·07	0·13	0·43
3-rhamnoglucoside	0·54	0·23	0·54
3-diglucoside	0·43	0·27	0·54
3-triglucoside	0·31	0·33	0·51
3-rhamnodiglucoside	0·41	0·34	0·61
7-rhamnoside 3-rhamnogalactoside	0·40	0·54	0·75
7-monoglucoside	0·54	0·02	0·17
7-monorhamnoside	0·75	0·02	0·18
3-xyloglucoside	0·55	0·29	0·65
3-monoglucuronide	0·53	0·67	
Quercetin			
3-monoarabinoside	0·70	0·07	0·31
3-monoxyloside	0·65	0·06	0·32
3-monoglucoside (isoquercitrin)	0·58	0·08	0·37
3-monogalactoside	0·55	0·09	0·35
3-monorhamnoside (quercetin)	0·72	0·19	0·49
3-rhamnoglucoside (rutin)	0·45	0·23	0·51
3-diglucoside	0·37	0·19	0·45
3-triglucoside	0·23	0·18	0·41
3-rhamnodiglucoside	0·36	0·26	0·54
7-monoglucoside (quercimeritrin)	0·37	0·00	0·07
4′-monoglucoside	0·48	0·01	0·13
3-monoglucuronide	0·40	0·69	
Myricetin			
3-monoglucoside	0·47	0·05	0·25
3-monorhamnoside	0·60	0·15	0·44
Gossypetin			
3-monogalactoside	0·42	0·10	
7-monoglucoside	0·31	0·02	

Table 16. *R_f's of flavonones, isoflavones, chalcones, aurones and their glycosides* (Harborne, 1959b)

	Butanol-acetic	Butanol-27% acetic acid (1/1)	Water	Acetic acid 30%
FLAVANONES				
naringenin	0·89	0·69	0·16	0·66
eriodictyol				0·56
hesperetin	0·89	0·85	0·11	0·67
naringenin 7-rhamnoglucoside	0·59		0·62	0·87
naringenin 7-monoglucoside	0·64		0·44	0·80
hesperetin 7-rhamnoglucoside	0·48		0·50	0·85
hesperetin 7-monoglucoside	0·60		0·33	0·79
ISOFLAVONES				
daidzein	0·92		0·08	0·62
genistein	0·94		0·04	0·59
genistein 7-monoglucoside	0·67		0·25	0·75
CHALCONES				
butein		0·83	0·01	0·19
okanin		0·56		0·08
butein 4′-monoglucoside		0·56	0·05	0·43
okanin 4′-monoglucoside		0·38		0·22
AURONES				
sulphuretin		0·80	0·01	0·19
aureusidin		0·57	0·01	0·10
leptosidin		0·67	0·01	0·19
sulphuretin 6-monoglucoside		0·49	0·03	0·41
aureusidin 6-monoglucoside		0·28	0·01	0·16
aureusidin 4-monoglucoside		0·49	0·02	0·25
leptosidin 6-monoglucoside		0·51	0·03	0·33

complex glycoside may be broken down no further than to a simpler one, which it is necessary not to mistake for an aglycone. The possibility has also been mentioned (3.4) of a flavanone being converted into a chalcone during the course of this hydrolysis. Lastly some pigments, e.g. myricetin, may be destroyed during an unduly prolonged hydrolysis.

After hydrolysis and cooling, the aglycone is extracted by a volume of ethyl acetate or isoamyl alcohol just sufficient to provide a supernatant layer. Using micro test tubes, it is possible to work with no more than a few drops of glycoside solution.

The aglycone, concentrated in the organic phase, is identified by chromatography (5.7) and by spectrophotometry in the visible and ultraviolet (5.11, 5.12). In principle, this identification is not difficult unless one is dealing with rare or novel substances. If it should seem to be necessary, the characterisation of the aglycone can be completed by the identification of the products of alkaline fission. The method is described later for the anthocyanidins (6.5). Under these conditions quercetin, like cyanidin, yields phloroglucinol and protocatechuic acid. The molecule can also be degraded by sodium amalgam (6.5).

5.9. *Examination of the sugars*

This must be done with a solution of the purified glycoside, especially one free from the natural sugars of the plant tissue (3.3).

Qualitative examination As in the case of the aglycones, this step requires complete hydrolysis of the glycoside. The aglycone is then extracted by ethyl acetate or isoamyl alcohol. The sugars remain in the strongly acid aqueous solution. It is necessary to concentrate this solution, and this requires the preliminary elimination of the hydrochloric acid.

This may be done with the aid of ion exchange resins (anionic resins Dowex 1 or Dowex 2). The author himself, however, prefers the method described by Harborne and Sherratt (1957), which consists in forming the chlorohydrate of di-*n*-octylmethylamine, which is extracted by chloroform, in which it is soluble. The sugars remain in the aqueous phase. The amine is used in 10% solution in chloroform, 30 ml of the reagent being needed to remove the acid entirely from 5 ml of 2N HCl. The method used (Ribéreau-Gayon, 1959) is as follows: 5 ml of the 2N HCl solution containing the sugars are shaken in a test tube for 2 min with 30 ml of the reagent. After standing, the upper layer is removed and the remaining solution filtered, because there are always floccules which settle with difficulty. The filtrate, collected in a 20×2 cm test tube, is evaporated to dryness *in vacuo*. The residue, dissolved in 2 drops of water is chromatographed on Whatman No. 1 paper with the upper layer of ethyl acetate–pyridine–water (2 : 1 : 2). This solvent, first employed by Forsyth and Quesnel (1957a), gives a good separation of glucose and galactose and of xylose and arabinose. Butanol–acetic acid may also be used for this separation.

There are several reagents that may be used to detect the sugars on the chromatogram (Lederer and Lederer, 1957), more or less specific for the different types, i.e. aldoses, ketoses, glucuronic acid, etc. A spray of aniline phthalate (phthalic acid 1·66 g, aniline 0·93 g, 95% ethanol 100 ml) is commonly used, the paper, after spraying, being heated for 5 min at 100°. Sensitivity of detection is increased by viewing in ultraviolet light. In the case of glucuronic acid, two spots appear, one due to the acid itself, the other to its lactone, the formation of which is favoured by treatment with acid.

In using this method, it is important to bear in mind that during the chromatographic operations and the elution necessary to isolate the pigment it is possible that sugars, or substances yielding sugars on hydrolysis, may be dissolved from the paper. The number and kind of such sugars will vary with the paper and the solvent employed. If the latter contain mineral acid the formation of sugar will be considerable, but even with nothing but organic solvents the effect can be observed. Great care must therefore be exercised and the precaution of carrying out a parallel treatment on a blank paper, and comparing the result with that of the experiment proper, is recommended.

Determination of the number of sugar residues When the nature of the sugar residues present has been determined, it may be desired to know how many such residues are attached to the aglycone. This will require quantitative analysis both of the aglycone and of the different sugars present. When dealing with microquantities of these substances the errors of determination are necessarily large, but since it is only necessary to decide whether there are one, two, or possibly a third sugar residue of a particular type present, an accuracy of the order of 10-20% is sufficient. Such methods, applicable equally well to anthocyanins as to flavones and flavonols, have been described by Harborne (1960) and by Chandler and Harper (1962).

After hydrolysis, the aglycone is determined spectrophotometrically (5.11, 5.12) by comparison with a pure specimen. The sugars are determined by elution of the spots revealed by spraying with *p*-anisidine or aniline phthalate and measuring the absorbance of the solutions so obtained in the spectrophotometer. The results are calibrated by chromatography and identical treatment of sugar solutions of known concentration.

5.10. *Identification of the glycosidic pattern*

The methods so far described may be adequate for the complete identification of, for instance, monoglycosides, but when more complex glycosides are involved, further information is necessary to determine the precise structure; whether, for example, in the case of an aglycone combined with two sugar residues these are in the form of a 3,7-diglycoside or a 3-bioside. Several methods will be described for doing this.

Thin layer chromatography on polyamide Egger (1961, 1964) showed that the R_f's of the different glycosides of kaempferol, quercetin and myricetin differ not so much on account of the nature of the aglycone as that of the sugar residues with which they are combined and their positions. The author himself (Ribéreau-Gayon, 1964b) showed that glucuronides have particular characteristics which enable them to be distinguished from other glycosides.

The procedure has been described earlier (5.6). The R_f's using aqueous ethanolic acetylacetone as solvent are as follows:

3-monoglucuronides	0·05
3-monosides	0·22
3-biosides	0·37-0·51
3,7-dimonosides	0·63
3-biosides-7-monosides	0·65-0·67

This method, which is simple to use, merits extension to all the known glycosides.

Controlled chemical hydrolysis As already mentioned, the ease of hydrolysis of glycosidic bonds depends on their configuration. Harborne (1965a) has determined the rate of hydrolysis of flavonoid glycosides at 100° in a medium consisting of equal volumes of 2N HCl and 95% ethanol. The results, collected together in Table 17, show that the nature of the aglycone has little effect on the rate of hydrolysis; the important factors are the nature of the sugar residue and the position of its attachment to the aglycone. L-Rhamnose and L-arabinose are the most easily detached, next D-glucose and D-galactose, and lastly D-glucuronic acid is especially difficult to separate from the aglycone. As regards position of attachment, 3-glycosides are most susceptible, next 4′-glycosides, and lastly 7-glycosides, which are the most resistant.

These data show that it is possible not only to establish the structure of a glycoside by the use of appropriate procedures, but also to prepare simpler glycosides from complex ones when these are more easily available. The latter possibility is especially useful in the case of

3,7-diglucosides of flavonols. Controlled hydrolysis will liberate the sugar residue in position 3, leaving the 7-glucoside which can be identified by paper chromatography. It will have a low R_f in dilute acetic acid, and in ultraviolet light will appear yellow in colour, while the 3,7-diglucoside will appear dark-brown (5.5).

Table 17. *Rates of hydrolysis of flavonoid glycosides, on heating at* 100°C *in* 2N HCl-EtOH (1 : 1) (Harborne, 1965a)

Glycoside	Time for 50% hydrolysis (min)	Approximate time for complete hydrolysis (min)
kaempferol 3-rhamnoside	0·72	2–3
quercetin 3-rhamnoside	0·72	
quercetin 3-galactoside	1·5	4–6
kaempferol 3-glucoside	1·8	
kaempferol 7-rhamnoside	2·0	
quercetin 3-glucoside	2·3	
quercetin 4′-glucoside	5·0	8–10
apigenin 7-glucoside	5·2	15
kaempferol 7-glucoside	7·8	20–25
naringenin 7-glucoside	7·8	
quercetin 7-glucoside	12	
luteolin 7-glucoside	16	30–60
kaempferol 3-glucuronide	19	
quercetin 3-glucuronide	35	
quercetin 7-glucuronide	88	180
apigenin 4′-glucuronide	98	
kaempferol 7-glucuronide	101	
apigenin 7-glucuronide	134	250

Controlled hydrolysis is carried out in 1% aqueous HCl for a time varying between 5 and 30 min. At this low concentration of acid, chromatography is not affected and the hydrolysate can be chromatographed directly, without removing the acid, provided the glycoside is sufficiently concentrated. This method does not work well with the

3-biosides of flavonols and therefore it is not possible to obtain the 3-monoside. For instance, rutin, the 3-rhamnoglucoside of quercetin, gives practically no 3-glucoside by controlled hydrolysis (Harborne, 1959b). On the other hand, a good yield of 7-monosides of flavones may be obtained by the controlled hydrolysis of their 7-biosides.

Enzymic hydrolysis Harborne (1965a) has shown that commercial β-glucosidase will hydrolyse β-linkages between glucose or galactose and an aglycone so long as these are present as monoglycosides. Biosides are not hydrolysed by this preparation. For instance the 3,7-diglucoside of quercetin is hydrolysed but rutin, the 3-rhamnoglucoside of quercetin, is not. β-glucuronidases exist which are specific for glucuronic acid linkages. The hydrolysis is carried out at 37°C in acetate buffer, pH 5·0, with concentrations of glycoside and enzyme in the order of 1 mg per ml. The hydrolysis takes less than one hour.

Determination of the position of sugar residues The first step is complete methylation of the glycoside, for instance with methyl sulphate. After hydrolysis a partially methylated aglycone is obtained, the only free hydroxyls being those previously combined with sugar residues. Thus quercetin 3,7-diglucoside gives, after hydrolysis, 3′,4′,5-tri-*O*-methylquercetin. The product is compared chromatographically and spectrophotometrically with known reference compounds. The need for these substances, which have to be synthesised, limits the usefulness of this method. In this connection, the work of Nordström and Swain (1953) and Harborne and Geissman (1956) should be consulted.

Other methods have been reported, which may be used to confirm or amplify the results obtained with the preceding ones. Chandler and Harper (1961) have described controlled oxidation techniques using hydrogen peroxide, potassium permanganate or ozone, which give indications of the manner in which the sugar residue is combined with the flavonoid aglycone. For instance, hydrogen peroxide breaks only the glycosidic bonds in the 3-position, this reaction being described for anthocyanins (6.5). It will be seen (5.11) that the position of sugar residues may be deduced from the spectrophotometric behaviour.

Spectrophotometry

5.11. *Spectrophotometry in the visible and ultraviolet of flavones and flavonols*

Many studies have been made of the spectrophotometry of flavonoid pigments in the visible and ultraviolet, a most valuable method of identification of these compounds. Reviews on the subject have been published by Geissman (1955), Jurd (1962a) and Harborne (1964d). The procedures for measuring absorption spectra have already been described (3.16).

The limiting structures involved in the mesomeric forms of the flavones (R=H) and the flavonols (R=OH) are represented by (31), (32) and (33). They correspond with different localisations of the free electrons (1.2). These bodies absorb in two regions of the ultraviolet spectrum, the first between 320 and 380 nm (band 1) and the second between 240 and 270 nm (band 2). Jurd considers that band 1 corresponds with the absorption of form (32) which involves the conjugation of the CO group with the B ring, band 2 being due to the structure (31).

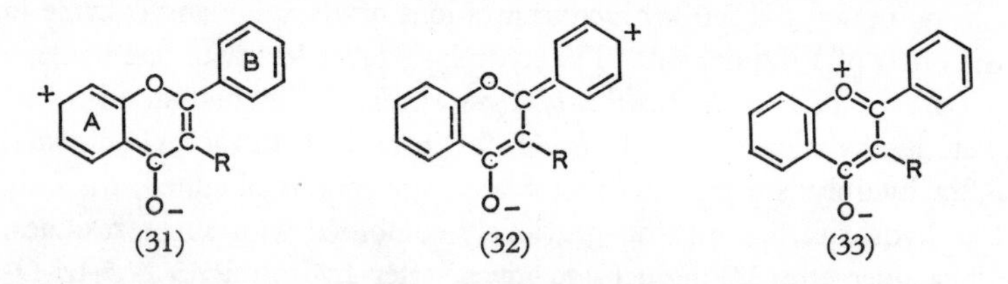

This interpretation depends on the effect on absorption of substitution of the A and B rings (Table 18). Substitution of an OH group in the B ring has a greater effect on band 1 than has substitution in the A ring, the effect being converse for band 2.

Table 18. *Influence of substitution in the A and B rings on the spectrum of the flavonols* (Jurd, 1962a)

	Band 1 (λ_{max}, nm)	Band 2 (λ_{max}, nm)
Flavonol (unsubstituted)	344	239
5,7-Dihydroxyflavonol (substituted in A)	360	267·5
3′,4′-Dihydroxyflavonol (substituted in B)	368	250
5,7,4′-Trihydroxyflavonol (substituted in A and B)	367·5	266

The absorption maxima of the principal members of these two classes are given in Table 19. Figs 10, 11 and 12 show that the flavones may have a subsidiary peak in the form of a shoulder at 300-310 nm which may be useful for their identification.

Table 19. *Absorption maxima of flavones, flavonols and their glycosides*

	Band 1 (λ nm)	Band 2 (λ nm)
apigenin	336	269
luteolin	350	255
tricin	348	270
kaempferol	367·5	268
kaempferol 3-monoglucoside	350	267
kaempferol 7-monoglucoside	368	268
kaempferol 3,7-diglucoside	350	267
kaempferol 3-monoglucuronide	352	264
quercetin	375	257
quercetin 3-monoglucoside	360	258
quercetin 7-monoglucoside	375	257
quercetin 3,7-diglucoside	360	258
quercetin 3-monoglucuronide	364	258
myricetin	378	255
myricetin 3-monoglucoside	365	252
isorhamnetin	369	254
quercetagetin	361	272–253
gossypetin	386	278–262
fisetin	370	315 252

Consideration of the absorption spectra of aglycones and their glycosides leads to the following conclusions:

(*a*) Hydroxylation of the molecule produces a bathochromic shift (i.e. towards a longer wavelength) of band 1 and a hypsochromic shift (towards a shorter wavelength) of band 2.

(*b*) Methylation and glycosylation of OH groups have similar effects, resulting in a hypsochromic shift of band 1. As a rule, band 2 is little affected. This shift of band 1 varies according to the position of the OH group involved in the glycosidic linkage (Table 20).

Table 20. *Effect of glycosylation of the band 1 absorption of quercetin and its glycosides* (Harborne, 1964d)

	Band 1 (λ nm)	Δ λ (nm)
quercetin	375	
quercetin 7-monoglucoside	374	− 1
quercetin 7,4′-diglucoside	371	− 4
quercetin 3′-monoglucoside	367	− 8
quercetin 4′-monoglucoside	366	− 9
quercetin 3,7-diglucoside	363	−12
quercetin 3-monoglucoside	360	−15
quercetin 3,4′-diglucoside	350	−25

(*c*) The spectra are not affected by the nature of the glycosidic residue (galactose, arabinose, rutinose, glucuronic acid, etc.) except in the case of rhamnose. The 3-rhamnosides of the flavonols have their peak of absorption (band 1) displaced about 10 nm towards the shorter wavelength compared with the corresponding 3-glucosides.

Most flavonoids have a free hydroxyl group in position 5 which is able to form, with the CO group, an aluminium complex when this element is added in the form of aluminium chloride (2.4). The OH in position 3, when not engaged in glycosidic formation, is also able to form a complex. Complex formation is accompanied by a strong bathochromic shift (50 nm) of the whole spectrum, sometimes with duplication of bands 1 and 2 (Figs 10 and 11). The OH groups in positions 3 and 5 do not behave identically. In the case of an OH in position 3, the complex is more stable and the shift larger (60 nm) than with one in position 5 (40 nm) (Figs. 10 and 11, a comparison of apigenin and kaempferol).

Valuable information may also be obtained from complex formation with boric acid (3.16) (Jurd, 1956). In this case, complex formation only takes place when there is a free *ortho*-dihydroxy group. Apigenin, for instance, does not react, but luteolin does. Complex formation then results in a bathochromic shift of band 1 of 15 to 30 nm. *Ortho*-dihydroxy groups can also be detected by making use of the instability of their $AlCl_3$ complex in presence of HCl (2.4, 3.16). On addition of $AlCl_3$, complex formation takes place with *o*-diOH groups, and with the CO group in conjunction with OH groups in positions 3 or 5. If

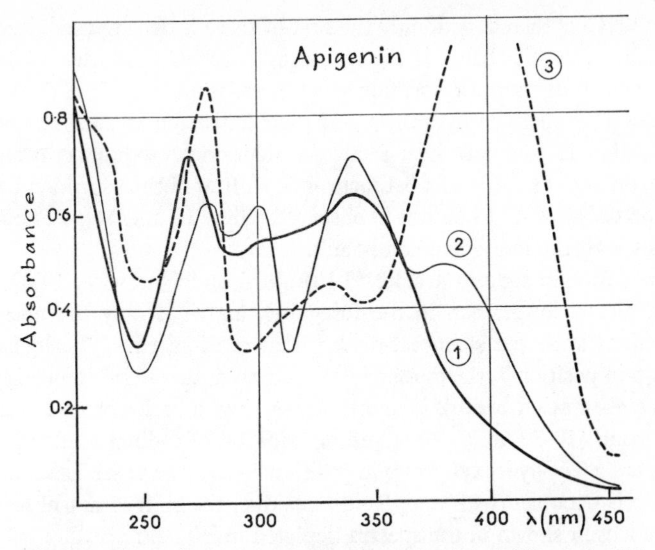

Fig. 10. *Absorption spectra of apigenin. 1. In 95% EtOH. 2. In 95% EtOH cont. 0·1% $AlCl_3$. 3. In 95% EtOH-0·002M NaOEt*

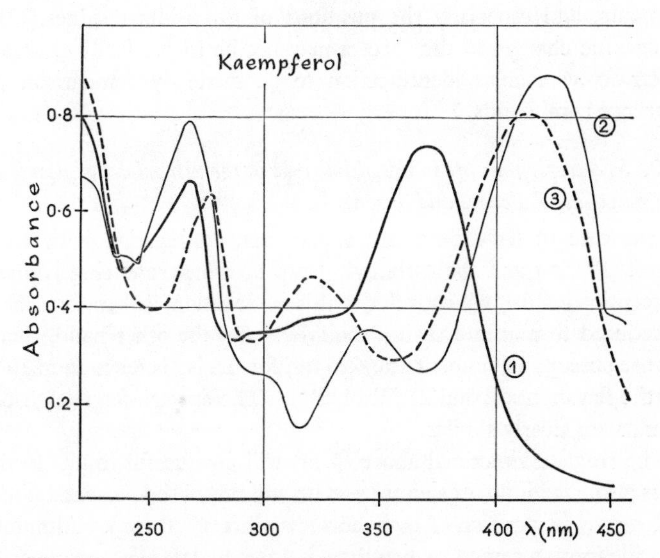

Fig. 11. *Absorption spectra of kaempferol. 1. In 95% EtOH. 2. In 95% EtOH cont. 0·1% $AlCl_3$. 3. 95% EtOH-0·002M NaOEt*

dilute HCl is then added, only the first of these is decomposed, leading to a characteristic change in the spectrum (Markham and Mabry, 1968).

When studying the absorption spectra, use is often made of the acidic function of phenolic hydroxyls and their ionisation in alkaline media. Ionisation is complete in a 0·002 M solution of sodium ethylate in 95% ethanol (3.16), and a bathochromic shift of both absorption bands of the order of 40 to 60 nm is observed (Figs. 10 and 11). The shift varies with the individual compound.

In alcoholic sodium acetate (3.16) (Jurd and Horowitz, 1957), the alkalinity is lower than in the foregoing, and it is only the phenolic functions with the strongest acidity that are ionised. The hydroxyl group in position 7, the most acidic, is affected; that in position 5 is not. Since absorption band 2 is due to the A ring, a bathochromic shift of this band (10-20 nm) is observed on addition of sodium acetate if this ring has a free hydroxyl group in position 7. On the other hand, if this group is occupied with a glycosidic residue, the shift is not observed. This is well shown in the spectra depicted in Fig. 12.

Some phenolic compounds are unstable in alkali. Myricetin especially is decomposed in sodium ethylate: a blue-violet colour changing to red is produced.

These reactions are often used to determine the structures of phenolic pigments, and especially the positions of glycosidic residues. Even though the changes in the spectra may not be theoretically explicable, spectroscopy enables identification to be made by comparison with reference compounds.

5.12. *Spectrophotometry in the visible and in the ultraviolet of flavanones, isoflavones, chalcones and aurones*

In the case of flavanones and isoflavones, conjugation between the lateral (B) ring and the carbonyl group of the pyrone ring is absent. Absorption band 1, which is due to this conjugation, is therefore lacking, or reduced to no more than a shoulder. On the other hand, there is intense absorption at about 250-270 nm for the isoflavones and 270-290 for the flavanones (band 2) (Table 21). The spectra for related compounds are clearly similar.

The reagents described above (5.11) will give useful results in these cases also. Addition of sodium acetate will reveal the presence of a free OH group in position (7 bathochromic shift of 10 nm). Aluminium chloride forms a complex when there is a free hydroxyl group in position 5 (bathochromic shift of 10-15 nm). Addition of sodium ethylate to

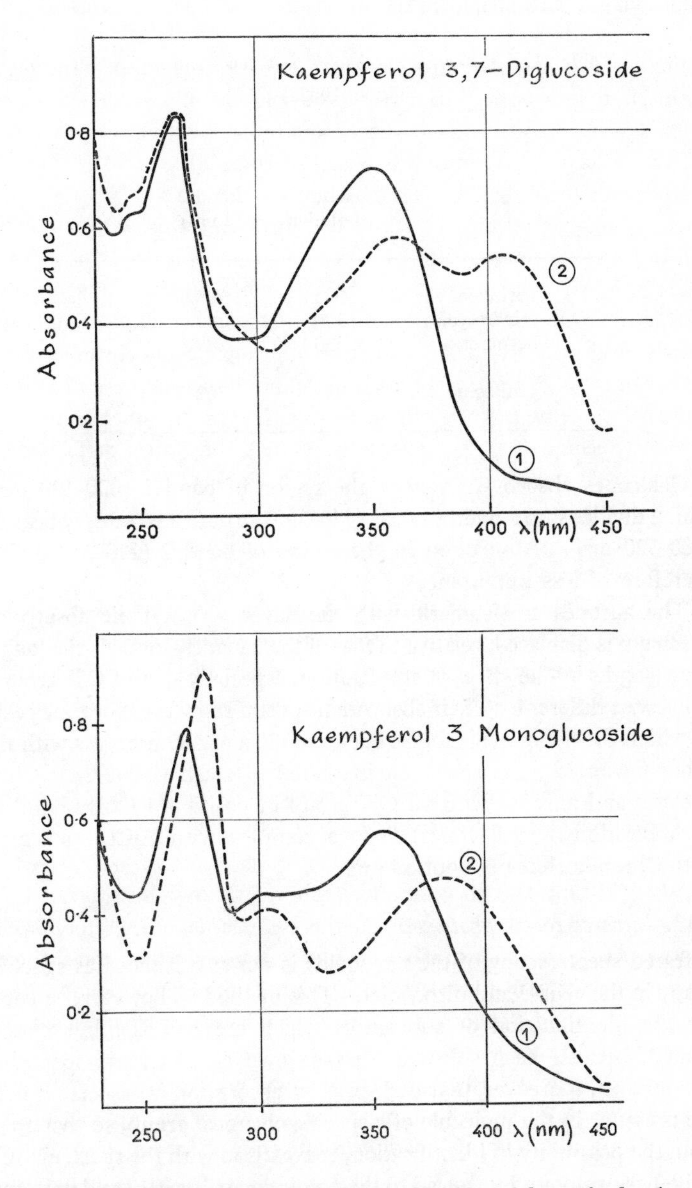

Fig. 12. *Absorption spectra of two glycosides of kaempferol. 1. In 95% EtOH. 2. In EtOH-NaOAc*

flavanones may cause the formation of chalcones, which will produce an intense bathochromic shift (3.4).

Table 21. *Absorption maxima of flavanones and isoflavones* (Jurd, 1962a)

	Band 1 (λ nm) shoulder	Band 2 (λ nm)
naringenin	325	288
eriodictyol	330	289
hesperetin	330	289
genistein	331	262

Chalcones absorb strongly in the region of band 1 (300-400 nm), which divides into a main peak 1a (340-390 nm) and a secondary peak 1b (300-320 nm). Absorption in the region of band 2 (220-270 nm) is therefore of less importance.

The aurones are isomeric with the flavones, but their absorption spectrum is displaced, relative to that of the flavones, towards the longer wavelengths. The effect of substitution, especially that of OH groups, is likewise different. Their absorption spectra show three or four peaks of which the main one is found between 380 and 430 nm. As with the other flavonoids, the spectrum is altered in alkali (0·002 M sodium acetate) and aurones with an OH group in position 4 (corresponding with position 5 in flavones) form a complex with $AlCl_3$, causing a bathochromic shift of about 60 nm.

5.13. *Infrared spectrophotometry*

Infrared spectroscopy of the flavonoids is not so advanced as spectroscopy in the visible and ultraviolet. This method is, however, valuable for the identification of substances which are complex but closely related, because each substance has its own characteristic spectrum. It is known, moreover, that each peak of absorption is associated with the presence in the molecule of a specific chemical group, so that apart from the possibility of identification, comparison with the spectrum of a reference compound may lead to the prediction of, or help to determine, the structure of an unknown constituent.

As a rule, the infrared spectra are measured in solution in a transparent solvent such as carbon tetrachloride or disulphide, but flavonoids are not very soluble in these solvents, so that it is necessary to work with dispersions of the product in Nujol (highly purified liquid paraffin), or better still in discs of potassium bromide which require lesser amounts of material (of the order of 1 mg). Either method gives the same result, at any rate in the case of the anthocyanins (Ribéreau-Gayon and Josien, 1960), but it is known that the spectra observed in the solid state do not always agree with those in solution.

In the case of potassium bromide, the substance, in crystalline form, is intimately mixed at the level of 0·5 to 1% with pure dry potassium bromide. The powder so obtained, compacted under high pressure, is transparent in the infrared. The spectrum is measured with a similar tablet of potassium bromide as a control. This method has the disadvantage that the substance has to be prepared in pure crystalline condition, which is not the case with ultraviolet spectroscopy. It is not, therefore, very suitable for use in conjunction with paper chromatography.

The major publications on the subject are those of Hergert and Kurth (1953), Henry and Molho (1955), Shaw and Simpson (1955), Inglett (1958), Briggs and Colebrook (1962), Lebreton (1962) and Wagner (1964).

It should be noted that infrared absorption data are expressed in wave numbers (ν) which are related to wavelengths by the expression:

$$\nu(\text{cm}^{-1}) = \frac{10^4}{\lambda(\mu)}$$

The following observations are drawn from Lebreton (1962):

(*a*) Free OH groups are responsible for absorption in the region of 3,300 cm^{-1}. When they are hydrogen bonded (i.e. in position 3 or 5) the frequency of this band is lower (3,100 cm^{-1}).

(*b*) Absorption due to CH_3 groups occurs at 2,930 cm^{-1}. It is often obscured by absorption due to OH groups (*a*).

(*c*) Absorption due to the CO group of the heterocycle occurs at 1,685 cm^{-1} in the case of flavanones, where the carbonyl group is conjugated only with the A ring. This is lowered to 1,650 cm^{-1} in the case of 5-hydroxyflavanones.

In the case of flavones, the carbonyl group is conjugated with both A and B rings. The absorption due to the CO group in

5,7-dihydroxyflavones is at 1,655 cm^{-1}, in 7-hydroxyflavones at 1,630 cm^{-1}.

In the case of flavonols, the carbonyl absorption is lowered by about 30 cm^{-1} compared with that of the homologous flavone.

(*d*) Aromatic double bonds absorb between 1,500 and 1,610 cm^{-1}, a second peak is present at 1,585 cm^{-1} when the double bond is conjugated with a benzene ring.

(*e*) Phenolic hydroxyl groups have strong absorption at about 1,360 cm^{-1}, with a secondary peak at 1,200 cm^{-1}.

(*f*) A peak at 1,165 cm^{-1} is due to *meta* dihydroxyl substitution (5,7-dihydroxy).

(*g*) Below 1,000 cm^{-1} the spectrum is difficult to interpret. At 800 cm^{-1}, however, absorption due to the hydrogen atoms of the benzene rings may be observed.

6

The Anthocyanins

Introduction

Like the flavones, chalcones or aurones, the anthocyanins undoubtedly constitute a special class of flavonoids. Indeed, particular attention has always been paid to these pigments of the plant kingdom, perhaps because of the beautiful red and blue colours that they impart to flowers and fruits in contrast with the dominant green of vegetation. The anthocyanins have been as much studied as the flavone pigments and such distinguished chemists as R. Willstätter, P. Karrer and R. Robinson have worked on their structures.

The first section in this chapter is devoted to a description of the different substances, aglycones and glycosides, which are found in nature. Since the anthocyanins have an ionic structure, giving them rather special chemical properties, a discussion of this is provided in the second section, together with an account of their visible, ultraviolet and infrared spectra. Special attention is given to their electron spin resonance spectra, which have been studied by Osawa and Saitô (1968).

The last two sections of this chapter describe the techniques of separation and identification by chromatography; the methods outlined are often substantially similar to those described in the preceding chapter, but specially adapted to the study of the anthocyanins. But, most frequently, an experimenter will wish to examine plant tissue for anthocyanins and other phenolic constituents at the same time; in such cases, the general principles of the techniques to be used have already been described in Chapter 3.

The principal substances

6.1. *The aglycones*

Six anthocyanidins are widespread in nature: pelargonidin (1), cyanidin (2), peonidin (3), delphinidin (4), petunidin (5) and malvidin (6); of these, cyanidin is the most common. All these compounds have the same basic structure, 2-phenylbenzopyrylium or flavylium (12a), and vary in the hydroxylation or methoxylation of the benzene rings; for example, pelargonidin is 3,5,7,4′-tetrahydroxyflavylium.

Besides these common structures, there are anthocyanidins of more restricted distribution which have the hydroxyl groups in the 5- or 7-position methylated (Harborne and Simmonds, 1964); these are hirsutidin (7-methylmalvidin), rosinidin (7-methylpeonidin), capensinidin (5-methylmalvidin) and finally 5-methyldelphinidin. These pigments have each been found at least once in plants, in particular in different species of the Primulaceae or Plumbaginaceae. Another rare pigment is aurantinidin (3,5,6,7,4′-pentahydroxyflavylium) which has been identified in the petals of *Impatiens aurantiaca* (Jurd and Harborne, 1968).

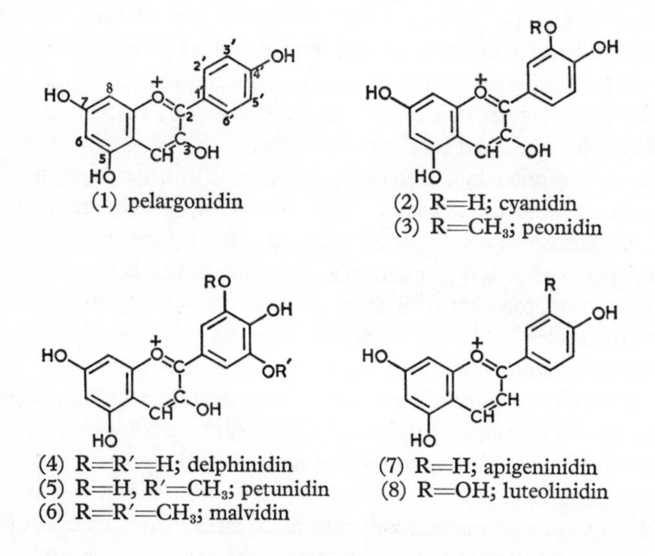

(1) pelargonidin

(2) R=H; cyanidin
(3) R=CH_3; peonidin

(4) R=R′=H; delphinidin
(5) R=H, R′=CH_3; petunidin
(6) R=R′=CH_3; malvidin

(7) R=H; apigeninidin
(8) R=OH; luteolinidin

In addition, there are four pigments which lack a hydroxyl group in the 3-position; the absence of this group affects their properties, since the 3-hydroxyl is the only one in the molecule which is not phenolic. These pigments are tricetinidin (5,7,3′,4′,5′-pentahydroxyflavylium), carajurin (6,4′-dimethoxy-5,7-dihydroxyflavylium) and also apigeninidin (7) and luteolinidin (8), which occur in the Gesneriaceae, in certain ferns (Harborne, 1966b) and in certain mosses (Bendz *et al.*, 1962).

In this account, we shall concentrate our attention on the six common anthocyanidins.

6.2. *The different glycosidic types*

As with the other flavonoids (5.2), anthocyanins do not occur in plants in the form of aglycones, but as glycosides; of the twenty different types known (Harborne, 1964b), the most widespread are the 3-glycosides. Except in the case of apigeninidin and luteolinidin, all known anthocyanins contain a sugar substituent in the 3-position. Glycosylation of this hydroxyl, which possesses special properties, is essential for the stability of the pigment (6.4). If, in the anthocyanin, a second hydroxyl is substituted by sugar, it is nearly always the 5-hydroxyl, so that 3,5-dimonosides are a very common type. There are a few rare exceptions, such as the 3-sophoroside-7-glucoside of pelargonidin found in the genus *Papaver*.

The sugar molecule most frequently encountered is glucose, but other sugars mentioned previously (5.2) also occur; of these, rutinose is found frequently. In addition, the presence of a branched trisaccharide (glucosylrutinose) attached to cyanidin in the 3-position has been reported. When a sugar is attached to the 5-position of an anthocyanidin, it is always glucose. Turning to the sugar, one finds that it is always attached to the aglycone through the one position. The glycosidic link is always β-, except in the case of arabinose, which is relatively rare in anthocyanins (Metche and Urion, 1961; Asen and Buden, 1966), when it is α-.

The principal glycosidic types are listed in Table 27 (6.13). Their distribution in nature has been studied by Harborne (1962b), who found the 3-monoglucosides, 3-rutinosides (3-rhamnosylglucosides) and the 3,5-dimonosides to be the most important.

The occurrence in plants of acylated anthocyanins has been known for a long time, well before corresponding derivatives of flavones and flavonols were reported. Generally speaking, the molecule of organic acid is esterified to the sugar hydroxyl and the phenolic hydroxyl groups are not involved. Harborne (1964a) studied fifteen acylated anthocyanins and found that the organic acid is always attached to the sugar molecule in the 3-position. In addition, this author identified only *p*-coumaric, caffeic or ferulic acids; acylation by sinapic, *p*-hydroxybenzoic or malonic acids, earlier reported in such pigments, could not be confirmed. Petanin, a common pigment in the Solanaceae, is an example of an acylated anthocyanin. Its structure (9) has been definitively worked out by Birkofer (1965; in reproducing it here, we have drawn the rhamnose linkage with the α-configuration). A similar derivative of delphinidin has been reported in the aubergine (Watanabé *et al.*, 1966).

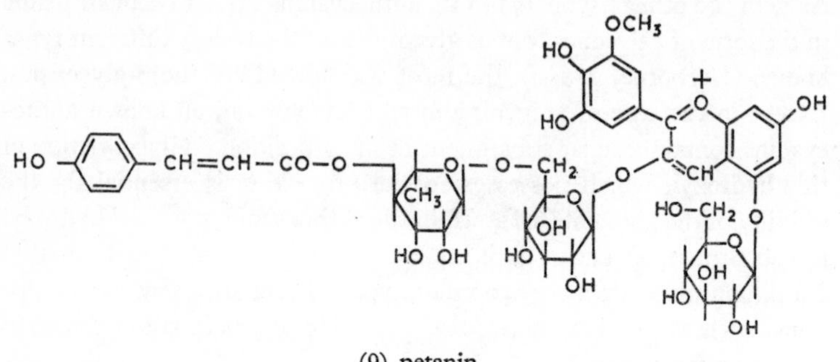

(9) petanin
(petunidin 3-(*p*-coumaroyl-4-rhamnosyl-6-glucoside)-5-glucoside)

In a recent paper, Albach *et al.* (1965) have shown that grapes (*Vitis vinifera* cv. Tinta Pinheira) contain delphinidin, petunidin and malvidin 3-monoglucosides acylated by *p*-coumaric acid which is joined to the 4-position of the glucose molecule, see e.g. malvidin 3-(*p*-coumaryl-4-β-D-glucoside) (10). A final example is cyanidin 3-(*p*-coumaryl-6-β-D-glucoside) which has been identified by Watanabé *et al.* (1966) in *Perilla ocimoides.*

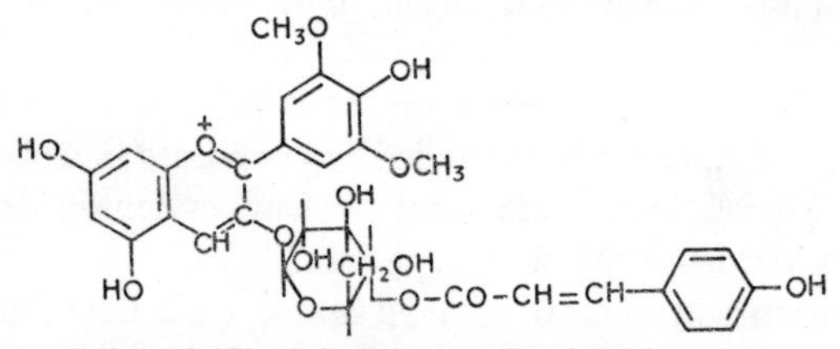

(10) malvidin 3-(*p*-coumaroyl-4-β-D-glucoside)

Structure and properties of the anthocyanins

6.3. *Structure of the anthocyanins*

The anthocyanidins are derivatives of 2-phenylbenzopyrylium or flavylium (12a); the pyrylium cation (11) is an oxonium ion, in which the tetravalent and positively charged oxygen atom has a structure similar to that of the nitrogen in the quaternary ammonium ion. Usually oxonium ions are less stable than ammonium ions; in the case of pyrylium with its 'aromatic character', it is the presence of the conjugated double bonds which gives it its stability (2.1) (Dean, 1963).

In view of the cationic character of flavylium and hence of the anthocyanidins, these compounds are always associated with an anion in the form of a salt; in the laboratory, anthocyanins are isolated usually as

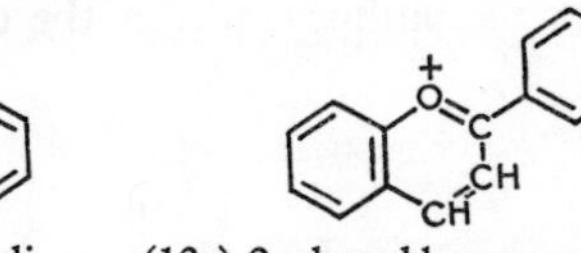

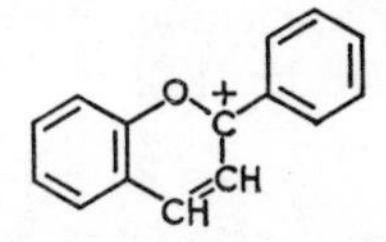

(11) pyrylium (12a) 2-phenyl benzopyrylium (flavylium), oxonium form (12b) 2-phenylbenzopyrylium, carbonium form

the chlorides, e.g. 'pelargonidin chloride' (13). In the plant, anthocyanins are associated with different anions (malate, citrate, etc.) and it is not possible to give an exact name to the molecule; one uses the words terms 'cyanidin' or 'cyanin' to refer simply to the cation in the molecule.

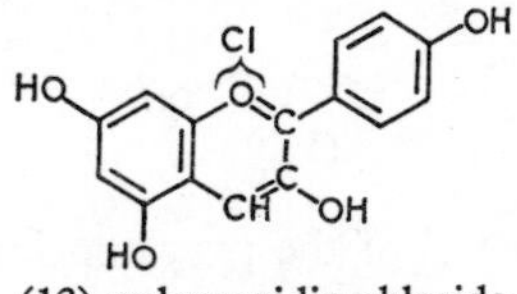

(13) pelargonidin chloride

Instead of considering the flavylium ion as an oxonium ion, it can be treated as a carbonium (or carbenium) ion, in which case the positive charge is located at carbon atom 2 which becomes trivalent (12b). The appearance of a positive charge on this carbon atom corresponds to the loss of one of the four outer shell electrons in the atom which can then only form three covalent bonds. Certain chemical properties of these compounds indicate there is a weak electron density around carbon atom 2 (Dean, 1963).

According to modern theories of electronic molecular structure, in a conjugated system such as the flavylium ion, the positive charge is not localised at one particular atom, but is spread over several atoms in the molecule. For flavylium and the anthocyanidins, one can write the formulae according to the oxonium structure (14a, 14b) and to the

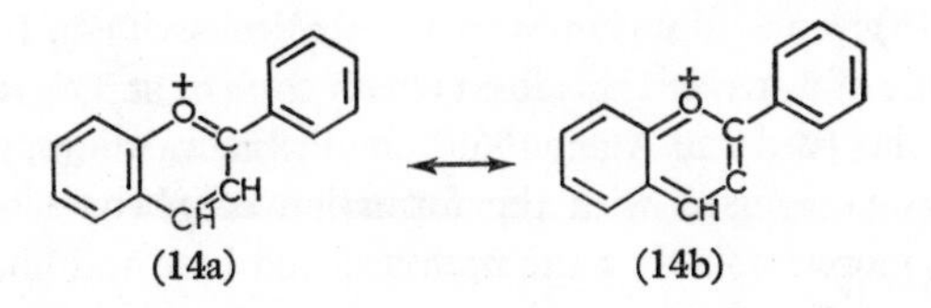

(14a) (14b)

carbonium structure (15a, 15b); in addition, in the case of anthocyanidins with hydroxyl substituents, the positive charge can be placed on one of these substituents (16a, 16b, 16c). Thus, the true structure corresponds to a 'resonance hybrid' between these different forms (2.1). In practice, most authors use either the oxonium (12a) or the carbonium (12b) representation. One usually writes the anthocyanins as the oxonium chlorides (13).

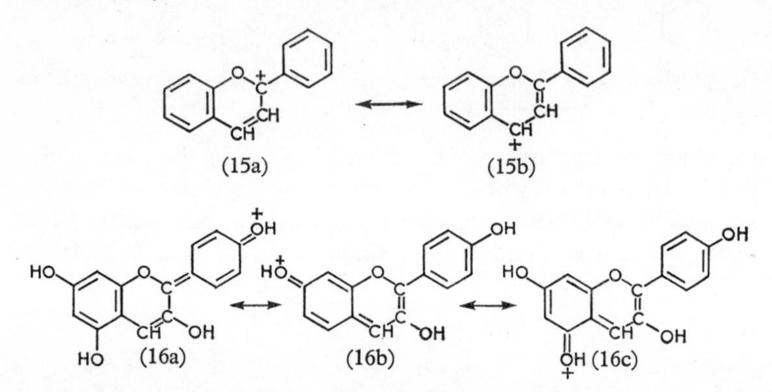

The 'aromatic character' of the pyrylium ion (11) gives it stability through the conjugated double bonds in the six-membered ring. However, the chemical properties of this ion are different from those of benzene; in particular, the pyrylium ion is not attacked by electrophilic reagents (low-electron-dense cationic groups) but is easily substituted in the 2-position by nucleophilic reagents (strong-electron-dense anionic groups) (6.5).

6.4. *Structural modification as a function of pH*

It is well known that the colour of anthocyanins varies with pH; red in acid medium, they become blue in neutral or alkaline solution. Moreover, the colour can disappear in a weak acid medium.

Taking as an example the case of cyanidin 3-glucoside, these transformations are interpreted in terms of the presence of several structures: the first corresponds to the red form (17); another, known as a pseudobase or carbinol base (18), is colourless and arises in weakly acid medium; (18) is in equilibrium with an anhydrobase (19a, 19b) with loss of a molecule of water and this form which absorbs at 538 nm is particularly unstable (Jurd and Asen, 1966); in alkaline medium, the phenolic groups become ionised with the formation of phenate or phenolate (20a, 20b), more stable than the undissociated base and blue in colour;

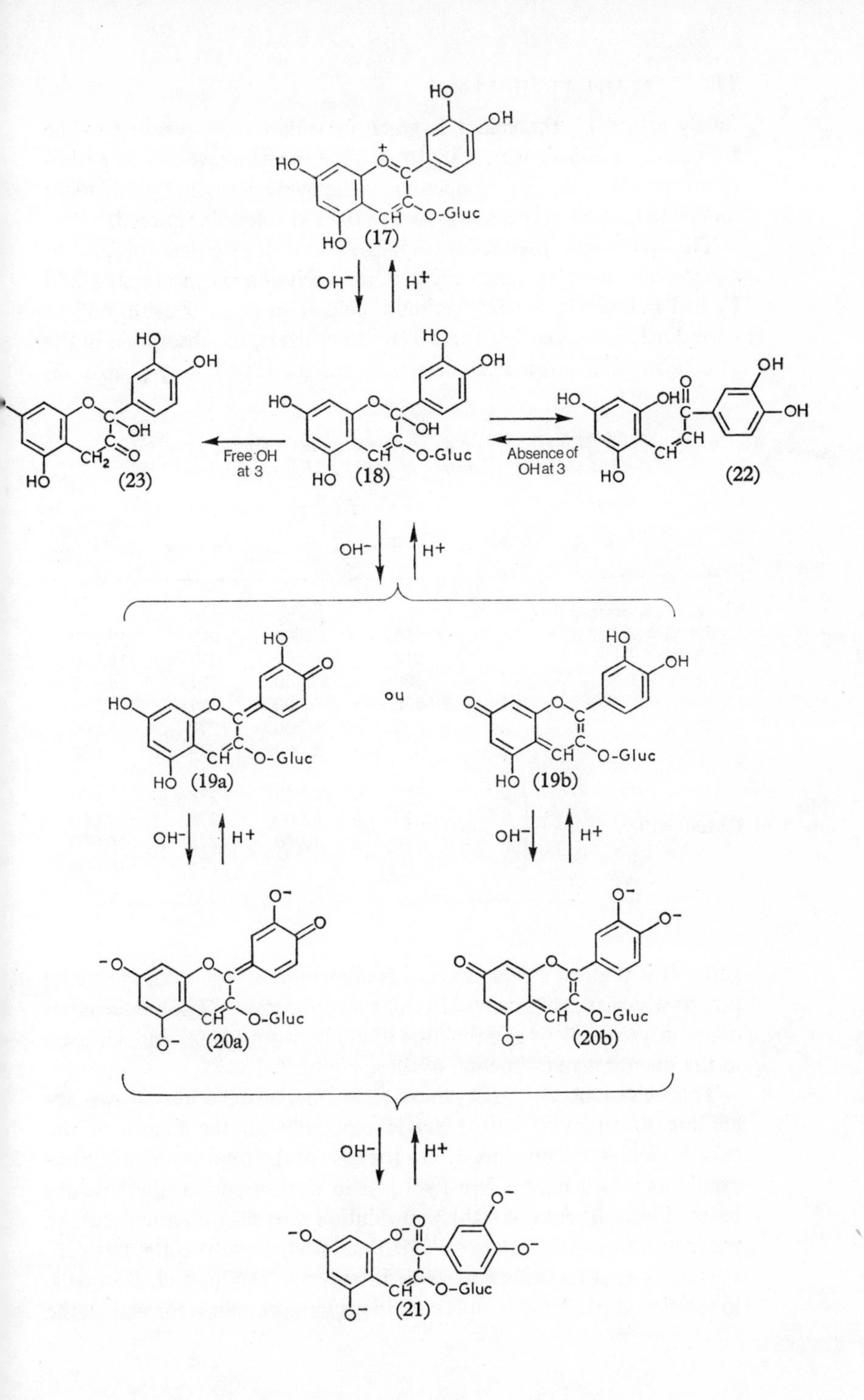

(17)
OH⁻
H⁺
(23)
Free OH at 3
(18)
Absence of OH at 3
(22)
O-Gluc
ou
(19a)
(19b)
(20a)
(20b)
(21)

finally at pH 12, the colour is green or yellow corresponding to the formation of a chalcone (21). In alkaline media above pH 8, anthocyanins are decomposed by opening of the pyran ring and by oxidation; on reacidification only a small part of the red colour is restored.

The colourless pseudobase appears at relatively low pH's. In aqueous medium, it is necessary to dilute a solution in concentrated HCl by half in order to obtain maximum coloration (λ_{max} 508-510 nm) of cyanidin 3,5-diglucoside (Table 22). In contrast, the absorption in the ultraviolet (277 nm) is not affected by variations in acidity (Table 22)

Table 22. *Absorption maxima and absorptivities of cyanidin 3,5-diglucoside as a function of pH* (P. Ribéreau-Gayon, 1959)

Milieu	pH	λ1 (nm)	ε1	λ2 (nm)	ε2
Conc. HCl, diluted to:					
90%		512	36,000	277	17,000
50%		510	35,000	277	18,000
10%		508	33,000	277	18,000
1%	1·0	508	30,000	277	17,000
0·1%	2·0	508	20,000	277	17,000
0·01%	2·9	510	5,000	279	16,000
Citrate buffer	2·4	510	12,000	278	17,000
	2·9	510	6,000	278	17,000
	3·9	510	1,000	278	16,000
	4·9		0	278	16,000

and this is understandable since this absorption is due to the phenolic groups which are not involved in this transformation. The modification of anthocyanin colour as a function of pH between pH 0·5 and 4 is used in the quantitative estimation of these pigments (2.15).

The behaviour of anthocyanins as a function of acidity varies according to structure and depends especially on the nature of the substitution at carbon atom 3. In the case of the most common anthocyanidins which have a free hydroxyl in the 3-position, the anhydro bases (19a, 19b) are not stable; on dilution they take up a molecule of water and form pseudobases. For their part, these pseudobases (18) are unstable, again being less stable in aqueous than in alcoholic media; in solution at pH 3, favourable conditions for pseudobase formation, the colour of the anthocyanidins progressively disappears and, after a certain length of time, the red form cannot be recovered by regeneration with acid (Nordström, 1956).

With a glycoside in which the sugar molecule is attached to the 3-position, the reaction is different (Ribéreau-Gayon, 1959). The colourless form (18) is stable and re-acidification regenerates the red form completely (17); this explains why 3-glycosylation of all the naturally occurring anthocyanins is necessary for their *in vivo* stability. Attachment of sugar in the 3-position is thus probably the final stage in anthocyanin synthesis (8.7). The author has sought (P. Ribéreau-Gayon, 1959) an interpretation of this difference in behaviour between the aglycones and their glycosides, due to the intervention of the 3-hydroxyl; it is assumed that in the case of the aglycones alone this enolic hydroxyl is irreversibly converted to a colourless ketonic form (23).

In contrast, if anthocyanidins lack a 3-hydroxyl group (apigeninidin (7) and luteolinidin (8)), the anhydrobases (19a, 19b) are stable and do not, on simple dilution, absorb a molecule of water to regenerate the pseudobase; acidification gives the red form unchanged (17). Certain anhydrobases have been identified in plants and they always correspond in structure to anthocyanidins lacking a 3-hydroxyl (Dean, 1963).

Furthermore, Jurd (1963, 1964) has shown that anthocyanidin pseudobases (18) undergo opening of the central heterocyclic ring only if there is a 3-hydroxyl present; this explains the formation of a chalcone with maximum absorption about 370 nm. This reaction which is complex, has been studied by Kuhn and Sperling (1960); Timberlake and Bridle (1965) have shown that light is involved. All these reactions have also been studied by Harper and Chandler (1967a, b) with the aid of different synthetic flavylium salts and using a polarograph.

6.5. *Chemical properties*

The chemical properties of the anthocyanins have probably been as well studied as those of other phenolic compounds, but the properties most frequently quoted are the degradative reactions used to establish their structures (Karrer, 1928). Equally, the reactions leading to the chemical synthesis of anthocyanins have been mentioned in numerous reviews (e.g. Hayashi, 1962). The reduction of flavonols (quercetin, 24) to anthocyanidins (cyanidin, 25) is particularly well known; this transformation was first carried out by Willstätter and Mallison in 1915 using methanolic HCl and magnesium. On the other hand, Lavollay and Vigneau (1943) have succeeded in preparing cyanidin (25) by the oxidation of epicatechin (26). These reactions are mainly of theoretical interest since they do not take place in plant tissues; nevertheless, they do show that the different types of flavonoid constituent are interrelated. Certain properties given here are of special interest because they are used in identification and in quantitative estimation.

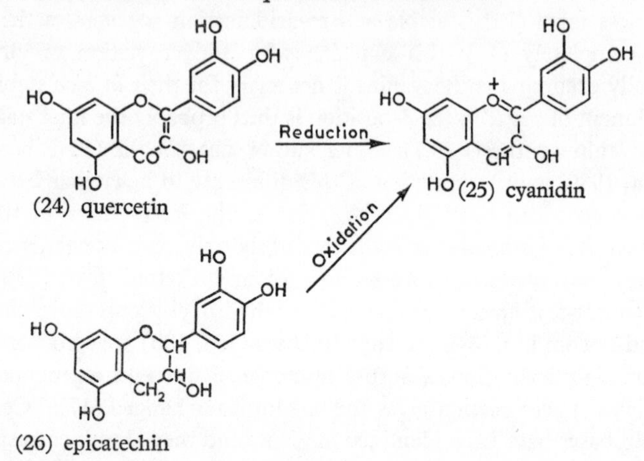

Reduction The fact that anthocyanins can be reduced is known but the mechanism of the reaction has not been studied. Sannié and Sauvain (1952) have put it as follows: 'reducing agents (zinc, aluminium or magnesium in acid solution) produce a rapid decolourisation of anthocyanin solutions which is permanent if the oxygen of the air is excluded. In the presence of oxygen, the colour is restored more or less quickly depending on the reaction conditions (i.e. nature of mineral acid employed).'

Zuman (1952) studied the polarographic reduction of the anthocyanidins. He observed a correlation between the polarographic inflections and the pH of the solution; at each change in colour, he noted a new inflection in the reduction curve, indicating the reduction of a corresponding structure. In tartaric acid buffer pH 3, the half-wave potentials are as follows:

pelargonidin	405 mV
pelargonidin 3,5-diglucoside	410 mV
cyanidin	400 mV
cyanidin 3,5-diglucoside	410 mV
delphinidin	425 mV
delphinidin 3,5-diglucoside	390 to 520 mV

Polarographic measurements on anthocyanins have also been carried out by Keith and Powers (1965).

In addition, Ribéreau-Gayon and Gardrat (1956) have shown that a solution of oenin (malvidin 3-monoglucoside) at pH 2·5 can be reduced by titanium chloride and then reoxidised by dichlorophenol-indophenol. They concluded that, even in the absence of catalyst, oenin constitutes a true reversible oxido-reductive system, the normal potential of which is about 300 mV at pH 2·5. In our present state of knowledge, little is known about the nature of the reduction product. This reaction, which causes decolourisation, must destroy the conjugation between the two benzene rings; one can assume at least that there is reduction of the double bond in the central pyran ring.

Jurd (1968b) has reduced different synthetic flavylium salts with zinc or magnesium in neutral or acid media and obtained dimeric products with 2-bis-flavene structures (e.g. 27). This reaction has not been tested with natural anthocyanins or anthocyanidins possessing a 3-hydroxyl group.

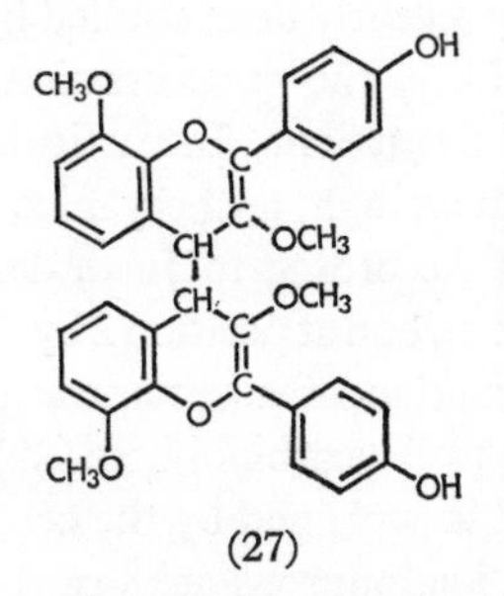

(27)

Action of Bisulphite The decolourisation of anthocyanins by sulphurous acid or alkaline bisulphite is also a very well-known reaction; it has been used as the basis of a method of quantitative analysis of the pigments (2.15). Several theories have been put forward to explain the reaction mechanism.

Genevois (1956) noted that anthocyanins react not only with bisulphite but also with other carbonyl reagents such as hydroxylamine, hydrazine and acetylhydrazide. We have shown that the reaction proceeds faster between pH 3 and 1 (P. Ribéreau-Gayon, 1959). Also it is reversible, but only in the case of anthocyanins; re-acidification or the addition of an excess of a carbonyl compound (acetaldehyde) restores the colour. These data are in favour of an addition reaction on the pseudobase, which does not involve the 3-position since the glycosides react in the

F

same way as do the aglycones. We have thus been led to propose that in a weak acid and in the presence of bisulphite, the colourless form (29) reacts in the form of the chalcone (30), which possesses a ketone function at position 2, to yield the combination (31).

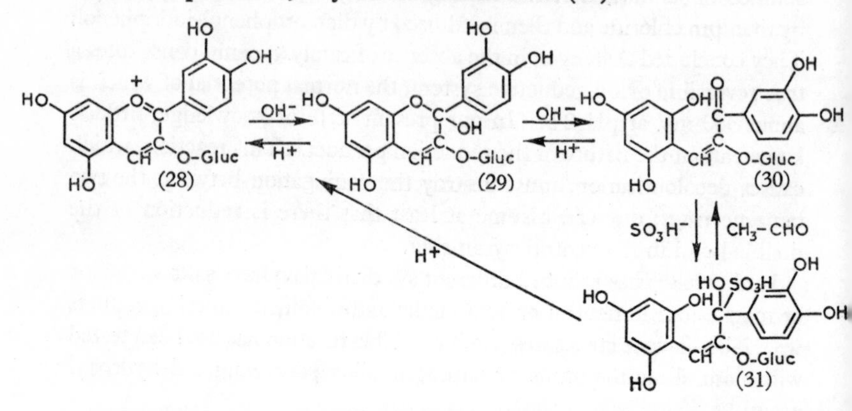

This problem has very recently been studied by Jurd (1964a) who has shown that, besides acidity, the concentration of the bisulphite ion controls the course of the reaction. In particular, if the concentration of bisulphite ions is high enough, anthocyanins can be decolourised in acid solution pH 0·63. According to Jurd, decolourisation is due to attack at the electrophilic carbon at position 2 by the nucleophilic HSO_3^-. This results in the formation of a chromene sulphonic acid, with a structure (32) analogous to the carbinol base (29). The slowness of the reaction in stronger acid is explained by the conversion of bisulphite to the less easily dissociated sulphurous acid, causing a diminution in the HSO_3^- ion concentration.

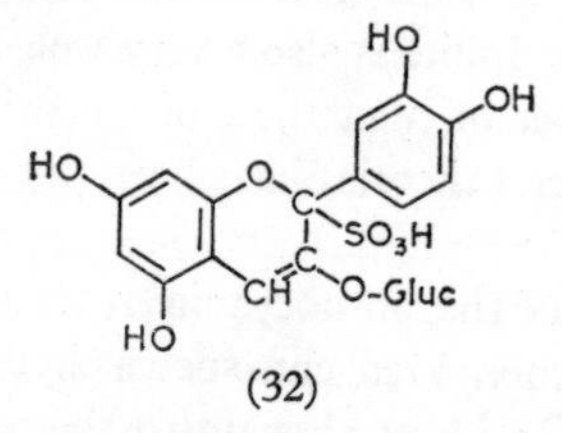

Jurd considers that the ketonic form (30) is not involved in the process, since the characteristic absorption of chalcones at about 370 nm has not been observed during the reaction; this agrees with our own observations. However, the mechanism which we have proposed does not require the presence of (30) in the free state. It is quite reasonable to suggest that such a structure is uniquely involved with the reactant to give immediately the addition compound (31) which lacks any special absorption bands in the ultraviolet, only having absorption due to the phenolic functions. In any case, the structures (31) or (32) proposed for the reaction product are very similar.

It is worth mentioning that anthocyanins react in a weak acid medium with potassium cyanide to give a green colour, possibly by a mechanism similar to that mentioned above.

Oxidation by hydrogen peroxide This reaction has been examined by Karrer and De Meuron (1932) for determining anthocyanin structure and especially for obtaining evidence for the presence of sugar residues in the 3-position. Oxidation of malvin (33) (malvidin 3,5-diglucoside) gives a crystalline product malvone (34), in which the sugar residue at position 3 is attached by an ester link; consequently, this molecule is saponified in alkaline solution, giving at the same time syringic acid, corresponding to the benzene B-ring; under these conditions the sugar molecule in the 5-position is not affected.

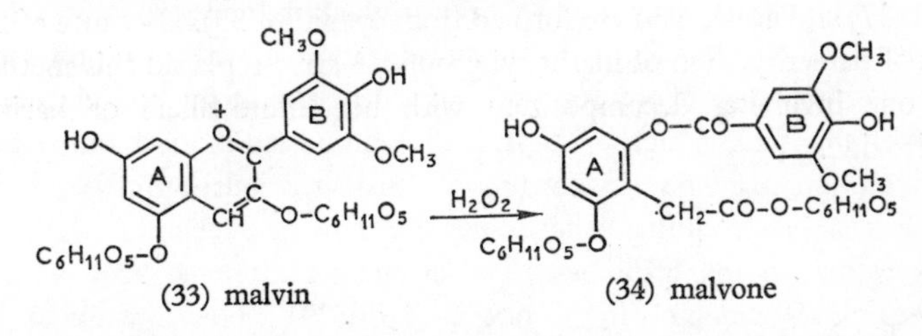

(33) malvin (34) malvone

This method has been used on a micro-analytical scale in conjunction with paper chromatography, for determining the nature of the 3-sugar in anthocyanins and other flavonoids (Chandler and Harper, 1961). These authors use the following technique: 1 mg of anthocyanin is dissolved in 0·2 ml methanol (or 1 mg of other flavonoids in 0·2 ml water containing 0·01 ml 0·1N NH_4OH); 40 μl 30% H_2O_2 are added and it is kept at room temperature for four hours; excess H_2O_2 is then decomposed by adding traces of palladium. The sugar in the 3-position is then hydrolysed by treatment for 20 hours with 50 μl 0·880 NH_4OH, the ammonia finally being removed by heating on a waterbath for 5 minutes. The solution so obtained is used directly for paper chromatographic identification of the sugar liberated.

The same method, applied by Harborne (1964a) to acylated anthocyanins, has allowed the isolation of the acylated sugars attached to the 3-position and moreover has shown that the acyl group is attached in this position. This reaction, used to decolourise the anthocyanins, has been used by Swain and Hillis (1959) as a method of quantitative estimation (2.15).

In addition, Karrer has shown that the aglycones (anthocyanidins) are easily destroyed by hydrogen peroxide; the quantity of the reagent formed by the action of ultraviolet radiation on water is sufficient to initiate the reaction; this explains the need for manipulating anthocyanidin solutions in the dark (Nordström, 1956).

The mechanism of this reaction depends on the conditions employed and it is possible to obtain other products, in particular 2-phenylbenzofurans (Jurd, 1964b). Hydrogen peroxide oxidation of a flavylium salt lacking a 3-hydroxyl (Jurd, 1968c) yields a C_6—C_3—C_6 derivative with diol and ketone functions; borohydride reduction of this compound gives a 2,3-*trans*-3,4-*trans*-flavan-3,4-diol.

Alkaline decomposition The alkaline fusion of anthocyanins was first used by Willstätter to determine the nature of the B-ring substitution pattern. Under these conditions, phloroglucinol (36) protocatechuic acid (37) and acetic acid are formed from cyanidin (35). Because of the risk of demethylation of methoxyl groups, Karrer replaced this method by one involving decomposition with hot dilute alkali or barium hydroxide.

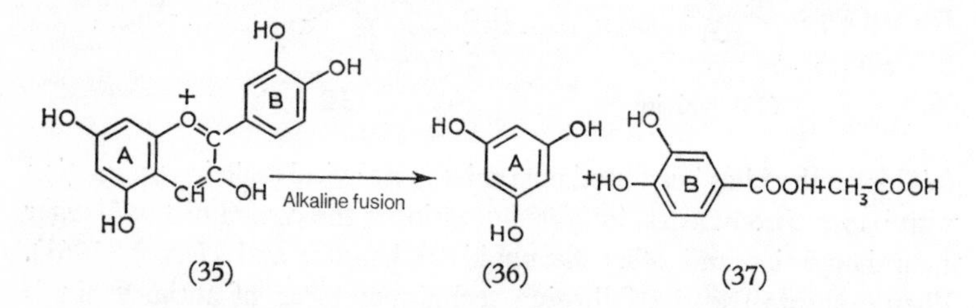

Although, in principal, chromatography is sufficient for the identication of the aglycones, alkaline fusion provides useful confirmation. One such procedure, applied on a micro-analytical scale to be used in conjunction with paper chromatography, has been described for the flavonoids by Bate-Smith and Swain (1953), Maurice and Mentzer (1954) and Masquelier and Point (1956). In applying this to the anthocyanins (P. Ribéreau-Gayon, 1959), it has been found that fusion and hydrolysis give similar results; if the conditions of fusion are not too drastic, the methoxyl groups are not affected. Nevertheless, gallic acid, formed from delphinidin, and methylgallic acid, from petunidin, are unstable and difficult to identify.

Our procedure is as follows: the anthocyanidin, produced from an anthocyanin, is usually in amyl alcohol (1 mg in about 5 ml); it can be transferred to the aqueous phase by shaking it several times with several ml. 1% HCl and an excess of benzene (4 to 5 volumes). This is evaporated to dryness under vacuum in a test tube; 200 mg KOH (ca. 2 pellets) and 4-5 drops of water are added. It is heated for several seconds in a blue bunsen flame; the original dark green colour becomes brown, then red-brown, then yellow; the heating is stopped before this final state. After cooling, 2 to 3 ml of water are added to it in a small test tube, and it is acidified with 6N HCl and the degradation products extracted into ether. The benzoic acids are then identified by paper chromatography (4.5, 4.6).

Reductive cleavage One of the problems of identification is that alkaline decomposition gives low yields and the degradation products are difficult to identify. Hurst and Harborne (1967) have outlined a method, useful for anthocyanins and above all for flavone pigments (5.8), involving reductive cleavage of the molecule with sodium amalgum. As before, the A-ring usually gives phloroglucinol, sometimes resorcinol and exceptionally another phenol; the B-ring gives a mixture of C_6—C_3 alcohol and acid with the same substitution. The authors used the following procedure: 20 ml 2N NaOH are placed in a 250-ml three-necked flask, fitted with a reflux condenser and a nitrogen gas flow. A preliminary heating at 95° removes any dissolved oxygen. After adding 15 g 2% sodium amalgum, heating is continued for several minutes to remove final traces of oxygen. The sample, if necessary as a spot cut from a chromatogram, is introduced while the nitrogen flow is maintained. After heating for 75 minutes, cooling at 20°, 5 ml concentrated HCl are added and the solution transferred to a 50-ml separating funnel, and extracted several times with 20 ml portions of ether; these extracts are concentrated to 1 to 2 ml.

Separations are effected by two-dimensional TLC on silica gel, the first solvent being 10% acetic acid in chloroform and the second 45% ethyl acetate in benzene. After the first development, the plates are dried for 10 minutes in a hot-air draught to remove the acetic acid. The phenols are detected with the Folin-Ciocalteau reagent, a variant of the Folin-Denis reagent (2.6), in the presence of ammonia vapour. Phenols having *ortho*-di- or trihydroxyls, are more easily separated on microcrystalline cellulose (Merck) using the solvents benzene–methanol–acetic acid (45 : 8 : 4) and 6% acetic acid in water. Identifications are made by comparisons with authentic samples; and by the fact that there is a relationship between chromatographic mobility and structure in the systems employed.

Condensation and polymerisation The possibility of the condensation polymerisation of anthocyanins or their condensation with other molecules has been mentioned frequently but there is very little definitely known on this point. The most important studies are those of Jurd and his collaborators at Albany (California), who have worked only with synthetic flavylium salts; hence their results cannot be related to the natural anthocyanins which all have a free hydroxyl in the 3-position.

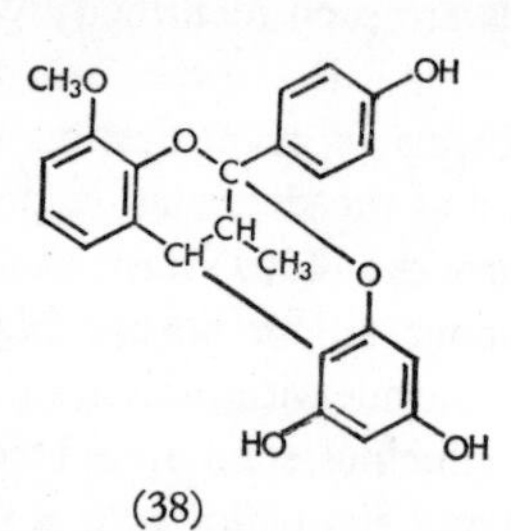

(38)

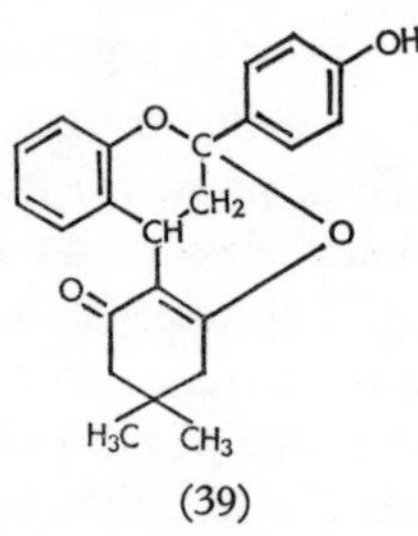

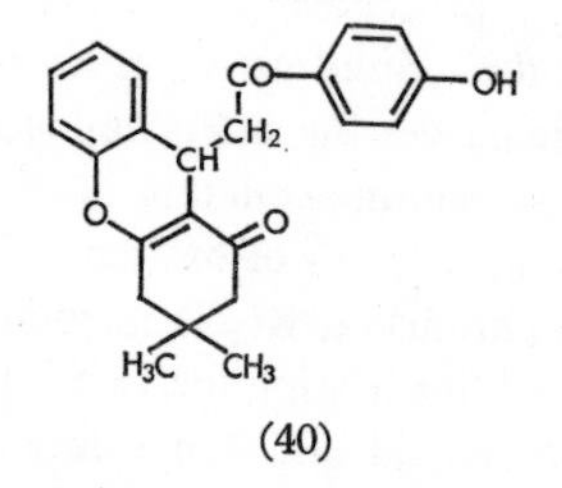

(39) (40)

The ability of flavylium salts to undergo dimerisation during reduction has already been noted (6.5); their ability to condense with catechins will be discussed in a later section (8.8); using these analogues, Jurd (personal communication) considers that flavylium salts can also undergo condensation with their colourless pseudobases. In a general way, flavylium salts which are electrophilic (2.7), react with nucleophilic reagents (6.3) (Dean, 1963); in particular, they condense with phloroglucinol (Jurd and Waiss, 1964) to give the condensation product (38). With 5,5-dimethylcyclohexane-1,3-dione, one obtains the derivative (39) at pH 5·8 (Jurd and Bergot, 1965) or an isomeric structure (40) in acid medium (Jurd, 1965).

Formation of metal complexes Anthocyanidins or anthocyanins with free *ortho*-dihydroxyl groups in the B-ring form complexes with iron or aluminium. In the absence of carbonyl groups, there is no possibility of chelation (2.4). One finds therefore that aluminium chloride gives a blue colour with *ortho*-diphenols only under weakly acid conditions. This allows one to separate cyanidin from peonidin on one side, and delphinidin and petunidin from malvidin on the other; the reaction can be carried out either on chromatograms or in solution.

The following technique may be used: 1 ml 95% ethanol are placed in one tube and 1 ml of an ethanol solution containing 6 g $AlCl_3$ per litre in another. To each tube one adds two drops of an amyl alcohol solution of the anthocyanidin to be identified. It is unnecessary to use a spectrophotometer since a visual comparison is sufficient. If the colours are both red, an *ortho*-diphenol is absent; a blue colour with $AlCl_3$ is unambiguous proof of the presence of such a compound.

The blue colours of anthocyanins in nature are due to the formation of complexes with magnesium, iron and even potassium; the role of aluminium in these natural complexes is controversial. In the naturally occurring blue pigments, the metal complex is frequently associated with polysaccharide (8.10).

6.6 *Visible and ultraviolet spectroscopy*

Spectroscopy of anthocyanidins The colour of anthocyanins varies according to the number and position of hydroxyl and methoxyl substituents; apigeninidin is yellow-orange, pelargonidin orange-red and delphinidin red-violet. Methylation of the B-ring hydroxyl groups has little effect on the absorption spectrum, so that we only need consider the spectra of pelargonidin, cyanidin and delphinidin. It is different with methylation of the 7-hydroxyl group; rosinidin and hirsutidin have different spectra from those of cyanidin and delphinidin (Table 23).

Fig. 13 illustrates the spectra of the three principal anthocyanidins in 95% ethanol containing 0·1% HCl (P. Ribéreau-Gayon, 1959). There is intense absorption in the visible which, as can be seen, is progressively displaced towards longer wavelengths in changing from pelargonidin (1OH) through cyanidin (2OH) to delphinidin (3OH). At the same time the absorptivity (ϵ) also increases. This absorption in the visible is due to the conjugation between the central heterocyclic system and the two benzene rings; it changes with the pH (6.4).

In order to obtain the details of the absorption curve between 300 and

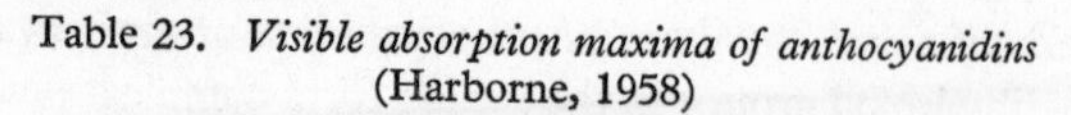

Table 23. *Visible absorption maxima of anthocyanidins* (Harborne, 1958)

	Solvent ethanol-HCl (nm)	Shift in the presence of $AlCl_3$ (nm)
apigeninidin	483	0
luteolinidin	503	+52
pelargonidin	530	0
cyanidin	545	+18
peonidin	542	0
rosinidin	534	0
delphinidin	557	+14
petunidin	558	+23
malvidin	554	0
hirsutidin	545	0

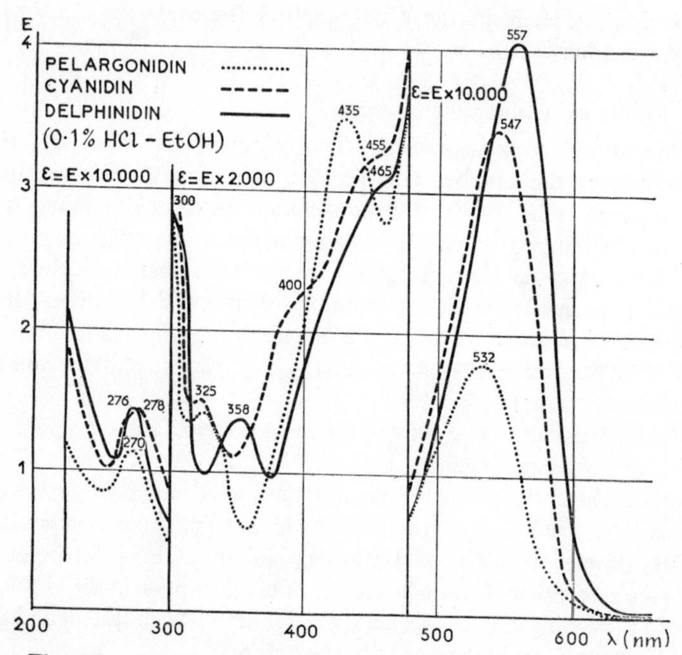

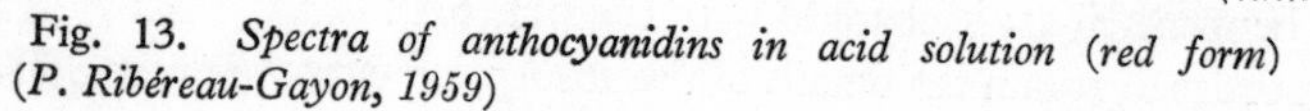

Fig. 13. *Spectra of anthocyanidins in acid solution (red form)* (*P. Ribéreau-Gayon, 1959*)

480 nm, which varies with the pigment, it is necessary to use more concentrated solutions. Finally, there is absorption in the ultraviolet between 270 and 280 nm, which corresponds to band 2 of the flavone spectrum; it is due to the presence of the phenolic groups in the molecule and is not affected by variations in pH, as long as the phenolic functions are not ionised (6.4).

The solvent plays an important but unexplained role in the position of the visible absorption, the remainder of the spectrum undergoing slight modification. In the case of the anthocyanidins, the maximal absorption in water is displaced by 25 to 35 nm towards shorter wavelengths as compared to the absorption in ethanol; in methanol, the hypsochromic shift is 10 nm.

As already mentioned (6.5), aluminium forms complexes only with anthocyanidins having free *o*-diphenol groups; this leads to a bathochromic shift in the visible of between 14 and 52 nm and there is a colour change to blue. In these conditions, the whole spectrum is altered (Fig. 14). Under strictly neutral alcoholic conditions, one obtains the spectrum of the blue form (Fig. 15).

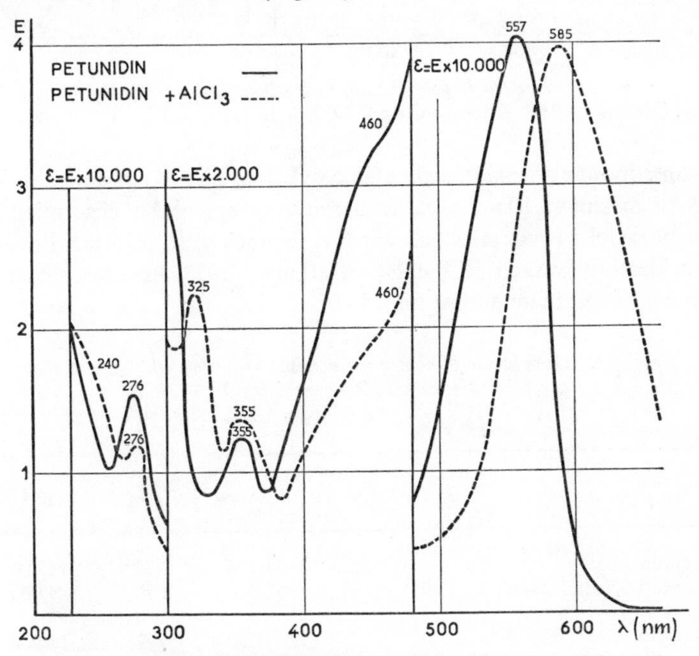

Fig. 14. *Effect of aluminium on the spectrum of petunidin.* (*P. Ribéreau-Gayon, 1959*)

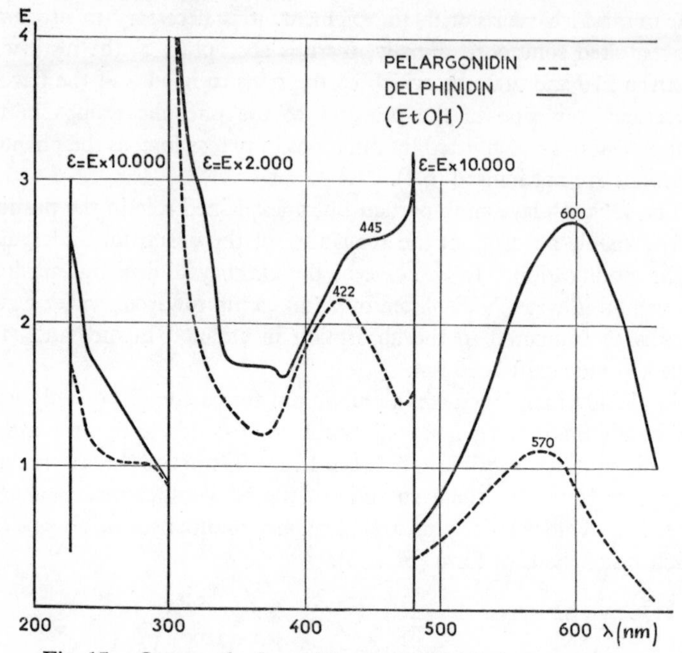

Fig. 15. *Spectra of pelargonidin and delphinidin in neutral solution (blue form). (P. Ribéreau-Gayon, 1959)*

Spectroscopy of anthocyanins Glucosylation shifts the position of the visible maximum 10 nm towards shorter wavelengths, but absorption in the ultraviolet is not affected. Finally, the molecular absorptivities are considerably lowered (Table 24). Harborne (1958) has shown that if the 5-position in an anthocyanin is free (3-glucoside), the spectrum of

Table 24. *Comparison of absorption maxima of anthocyanidins and their glycosides (in 95% EtOH containing* 0·1% *HCl)* (P. Ribéreau-Gayon, 1959)

	λ1	ε1	λ2	ε2
cyanidin	547	34,700	278	14,600
cyanidin 3,5-diglucoside	535	12,500	279	5,000
malvidin	557	36,000	277	13,300
malvidin 3-monoglucoside	545	10,300	278	5,400

the pigment has an inflection at 440 nm (Fig. 16). Thus, the ratios of the absorbances at 440 nm to those at the visible maximum for 3-glycosides are twice those for 3,5-diglycosides (Table 25).

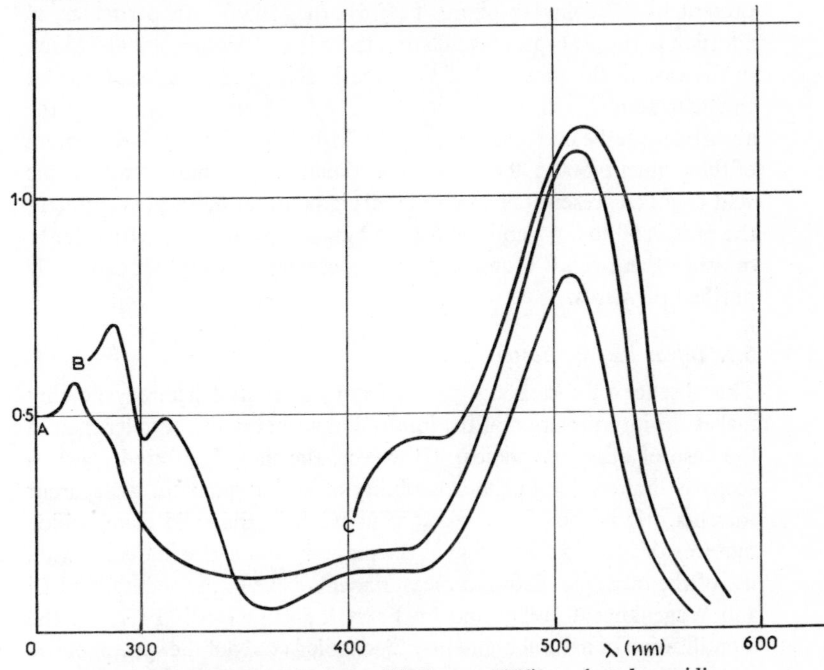

Fig. 16. *Anthocyanin spectra (Harborne, 1958). A. pelargonidin 3,5-diglucoside. B. pelargonidin 3,5-diglucoside acylated with* p-*coumaric acid. C. pelargonidin 3-monoglucoside*

Table 25. *Absorbance ratios of anthocyanins at* 440 *nm and at the visible maxima as a function of glucosylation at the* 5-*position* (Harborne, 1958)

	E 440/E max. (%)	
	5-OH free	5-OH substituted
derivatives of:		
pelargonidin	36–39	15–21
cyanidin, peonidin	19–26	12–13
delphinidin, petunidin, malvidin	16–19	9–11

Spectroscopy of acylated anthocyanins The spectra of acylated anthocyanins can be simply considered as the superimposition upon the spectrum of the simple anthocyanin of the spectrum of the acyl residues present in the complex pigment (Harborne, 1958). In particular, in addition to the absorption at 280 nm, there is a maximum at 308-313 nm in the case of the presence of *p*-coumaric acid, and at 326-329 nm for caffeic or ferulic acid substitution (Fig. 16). Moreover, the ratio of the absorbance at the maximum of the acid (310-330) to the visible maximum of the pigment (500-530 nm) is proportional to the number of cinnamic acid residues present per molecule of pigment; for one acyl substituent the ratio is 50 to 60% and for two acyl groups about 90%. In order to measure these ratios accurately, it is necessary to work with specially purified pigments.

6.7. *Infrared spectroscopy*

The absence of a carbonyl group from the central heterocyclic ring makes the interpretation of the infrared of anthocyanins simpler than in the case of other flavonoids. However, the use of infrared spectroscopy is limited due to the insolubility of anthocyanins in transparent solvents. It is therefore necessary to work in the solid state, which requires the previous isolation of the pigment and hence limits the wide use of the method. Infrared measurements have been used both by Li and Wagenknecht (1956) and by Forsyth and Quesnel (1957a) in the identification of natural pigments. A detailed study of the spectra of the six principal anthocyanidins has been carried out by Ribéreau-Gayon and Josien (1960).

These authors employed the solid state (5.13) either by dispersion in Nujol (purified paraffin oil) or preferably in the form of potassium bromide discs. The two methods gave similar results. In addition, the same authors examined two samples of malvidin in different crystalline form and obtained exactly similar spectra.

Although anthocyanidin spectra have been measured in the infrared region between 4,000 and 700 cm^{-1}, only the results obtained between 1,800 and 1,380 cm^{-1} will be discussed since these are the most interesting. The interpretation of these results have been covered in the following texts (Colthup, 1950; Bellamy, 1956; Jones and Sandorfy, 1956).

One observes in Fig. 17 the first well-defined absorption band at 1,637 $\pm$2 cm^{-1}, which is due to the conjugation between the heterocyclic double bond and the benzene ring. Between 1,600 and 1,500 cm^{-1},

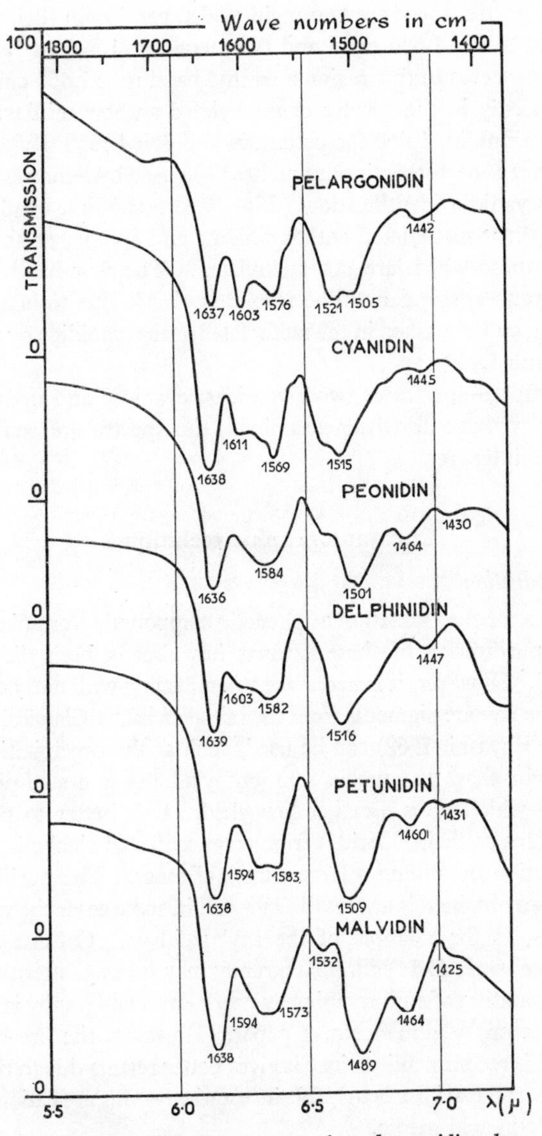

Fig. 17. *Infrared spectra of anthocyanidins between 1,400 and 1,800 cm⁻¹ (Ribéreau-Gayon and Josein, 1960)*

there are benzene ring absorption bands. Effectively, in all anthocyanidin spectra, there are two main peaks, varying in their intensity, the first at about 1,580 cm^{-1} and the second at 1,520 to 1,490 cm^{-1}. One also observes in this region a methyl band near 1,450 cm^{-1}. This is present only in the spectra of methylated anthocyanidins: at 1,464 cm^{-1} for peonidin, 1,460 for petunidin and 1,464 for malvidin. This band is a very characteristic one and has been used by Bendz *et al.* (1962) for anthocyanidin identification. Fig. 17 also shows a band at about 1,430 cm^{-1} for methylated anthocyanidins and another at about 1,445 cm^{-1} for those which are not methylated. The first band is due to methyl group absorption. The second is clearly due to benzene ring absorption and is masked in the methylated anthocyanidins by the bands at 1,460 and 1,430 cm^{-1}.

The infrared spectra of two glycosides, cyanidin and malvidin 3,5-diglucosides, have also been examined; the spectra are complex and difficult to interpret.

Chromatographic techniques

6.8. *Introduction*

Conditions for the extraction of phenolic compounds from plant tissues and the purification of these extracts has already been discussed in Chapter 3. However, it is necessary when dealing with anthocyanins to remove the flavone pigments from the initial extract. Classical chemical methods (Hayashi, 1962) can be used, such as the precipitation of the anthocyanins from an aqueous solution by adding saturated picric acid, a method which gives a rather low yield. It is better to precipitate anthocyanins as their chlorides from a concentrated alcoholic acid (1% HCl) solution by adding a large excess of ether. The purification of pigments so obtained is always difficult and in some cases there is partial hydrolysis, as for example of the acyl residues. Column (3.11) or preparative paper (3.14) chromatography may be used alternatively.

The separation of anthocyanins by paper chromatography is generally carried out on Whatman No. 1 paper. However, the use of Arches paper 302 is recommended since it gives better results due to the acidity of the paper; the acidity is helpful since anthocyanins are cations and are only stable in acid media.

As has been done for the other phenolic constituents, the special methods required for anthocyanins separation are now given. The reader is referred to Chapter 3 for details of the more general procedures used. It should be stressed that the beginner, as in the case of other phenolic constituents, should gain knowledge of all the aglycones liberated by acid hydrolysis before going on to study the glycosides.

6.9. *Solvents*

The ionic character of the anthocyanins and their much greater stability at low pH's requires the use of acid solvents, particularly for the aglycones (6.4).

Separation of the glycosides The best separations are obtained with butanol–acetic acid (*n*-butanol–acetic acid–water, 4 : 1 : 5) (5.4). It is, however, necessary when using this solvent to ensure that the extract to be chromatographed contains sufficient hydrochloric acid to keep the anthocyanins in the form of the chlorides during their separation.

As well as butanol–acetic acid, Harborne (1959a) has recommended the following solvents: butanol–hydrochloric acid (upper layer of *n*-butanol–2N HCl, 1 : 1); 1% aqueous HCl; acetic acid–conc. HCl–water (82 : 15 : 3); and 15% aqueous acetic acid. Fuleki and Francis (1967) suggest the use of the system *n*-butanol–benzene–formic acid–water (90 : 19 : 10 : 25), which can also be used for other phenolic compounds and for sugars.

For two-dimensional chromatography, the two pairs of solvents most frequently used are:

butanol–acetic acid and 15% acetic acid
butanol–hydrochloric acid and 1% hydrochloric acid.

Under these conditions, however, anthocyanins tend to decolourise and give large diffuse spots and this tends to limit the value of this particular technique. The author has obtained good results, however, in the separation on Arches 302 paper of the 10 to 15 pigments present in grapes by using as the first solvent the lower aqueous phase of the butanol–acetic acid mixture (5.4) and as second solvent the more usual upper phase (P. Ribéreau-Gayon, 1959). Since the first solvent is an aqueous one, the order of R_f's is different from that in butanol; but the fact that the solvent is saturated with butanol prevents the diffusion and decolourisation of spots that takes place in purely aqueous solvents. Derivatives of pelargonidin, cyanidin and peonidin separate better with less diffuse spots than do those of delphinidin, petunidin and malvidin.

Separation of aglycones Because of the considerable instability of the anthocyanidins (6.4), it is necessary to use very acid solvents and only a few are available. The solvent 'Forestal' (acetic acid–water–conc. HCl,

30 : 10 : 3) is certainly the best and has already been suggested for this purpose (5.4). Satisfactory two-dimensional separations have been obtained by the author using the Forestal solvent in conjunction with the butanol–hydrochloric acid solvent mentioned above; there is, however, some trailing of the anthocyanidin spots in the second solvent. Harborne (1959a) has suggested the use of butanol–acetic acid for the aglycones but it is necessary to prewash the chromatography paper with dilute hydrochloric acid.

6.10. *Chromatographic detection*

Since anthocyanins are coloured, their chromatographic detection is easy; in acid media, the colours are orange-red (pelargonidin), red (cyanidin and peonidin) or red-purple (delphinidin, petunidin and malvidin). With some experience, it is relatively easy to distinguish between pigments possessing one, two or three hydroxyl groups in the B-ring. There are also characteristic changes in these colours when the papers are fumed with ammonia.

In addition, examination in ultraviolet light is worth while since certain anthocyanins fluoresce brick-red in Wood's light; under normal conditions, the 3,5-diglucosides of pelargonidin, peonidin and malvidin (5-hydroxyl substituted and only one free hydroxyl in the B-ring) are the ones that are fluorescent. However, on strongly acid chromatograms, other diglucosides show some fluorescence. Finally, at low temperatures (−196°C) all anthocyanins become fluorescent and this increases the level of their detection (3.10).

Lastly, one can spray the paper with a 6% solution of aluminium chloride in ethanol and obtain blue colours for the anthocyanins which have two free hydroxyl groups in the B ring (6.5). In order to obtain a positive response, the acidity of the paper must not be too high and chromatographic solvents containing mineral acids should be avoided. Werckmeister (1965) suggests spraying successively with aluminium chloride solution and then, after drying, with acetate buffer pH 4·62.

6.11. *Column and thin layer chromatography*

Karrer and Strong (1936) and Karrer and Weber (1936) were the first to use column chromatography; this was for the separation of two anthocyanins on alumina and on calcium sulphate. Since these first experiments, the technique has been used several times by different workers, but it has not provided as good a resolution as that achieved by paper chromatography. Employing this method, Chandler and Harper (1958) separated seven anthocyanins from blackcurrant and Chandler (1958) two pigments from oranges. These authors used powdered cellulose—the most popular adsorbent for the purpose—and eluted the pigments with butanol–hydrochloric acid (6.9). The size of the cellulose particles, however, has a considerable effect on the success of the operation (Endo, 1957).

With regard to thin layer chromatography (5.6) (Paris, 1963; Nybom, 1964), the principal value is its speed of separation. As part of a general procedure for the analysis of plant tissues, Nybom (3.11) proposes two-dimensional chromatographic separations on thin layers of cellulose using the following solvents:

anthocyanidins
first dimension: HCl–formic acid–water (1 : 10 : 3)
second dimension: *n*-pentanol–acetic acid–water (2 : 2 : 1)
anthocyanins
first dimension: HCl–formic acid–water (4 : 1 : 8)
second dimension: *n*-butanol–HCl–water (6 : 4 : 2).

Methods of identification

6.12. *Introduction*

In practice, paper chromatographic techniques are used almost exclusively for anthocyanin identification. They have completely superseded the Robinson and Robinson tests of 1931 and 1932, which were based on the use of certain colour reactions to distinguish between the anthocyanidins and on partition between two immiscible solvents to separate the glycosides. For details of these tests, the reader is referred to Sannié and Sauvain (1952) and Hayashi (1962).

The chromatographic study of these substances requires the availability of reference compounds, a few of which can be obtained commercially. Anthocyanins can also be prepared by total synthesis or derived from other flavonoids. In addition, they can be isolated from authentic plant sources. These questions are dealt with in the review by Hayashi (1962).

6.13. *R_f comparison*

R_f values of the principal anthocyanins are collected together in Tables 26 and 27 (Harborne, 1959a, 1963). These values are only a general guide since it is not possible to make direct comparisons between

Table 26. *R_f's and colours of anthocyanidins* (Harborne, 1959a)

	R_f		
	Forestal	Butanol-acetic (paper washed with 1% HCl)	Colours
apigeninidin	0·75	0·74	yellow
luteolinidin	0·61	0·56	orange
pelargonidin	0·68	0·80	orange-red
cyanidin	0·49	0·68	red
peonidin	0·63	0·71	red
rosinidin	0·76		red
delphinidin	0·32	0·42	purple
petunidin	0·46	0·52	purple
malvidin	0·60	0·58	purple
hirsutidin	0·78	0·66	purple

experimental results and the literature data (5.7); a compound to be identified and the marker substance should always be chromatographed side by side on the same paper.

The relation between R_f and chemical structure (3.13) is especially interesting in the case of the anthocyanins since the number of possible aglycones is strictly limited. Measuring the R_f's of a small number of authentic glycosides allows one to estimate fairly precisely the R_f's of other pigments which are not available. This is particularly apparent when carrying out two dimensional chromatography, since the anthocyanins are arranged in order as a function of their structures (Fig. 6, Chapter 3). Nevertheless, R_f comparisons on natural mixtures of anthocyanins or anthocyanidins can only give an indication of the identity of the pigments present. A complete identification requires the isolation of each pigment in a pure state (3.14) followed by thorough analysis.

6.14. *Aglycone identification*

Acid hydrolysis of a glycoside is carried out by heating with 2N HCl for 30 minutes on a boiling water bath. There are no important differences

Table 27. *R_f's and colours of the principal anthocyanins* (Harborne, 1959a and 1963)

	R_f		
	Butanol-acetic	1% HCl	Colours
PELARGONIDIN			
3-monoglucoside	0·44	0·14	orange-red
3-monogalactoside	0·39	0·14	
3-rhamnoglucoside	0·37	0·22	
3-diglucoside	0·33	0·38	
3-xyloglucoside	0·34	0·30	
3-triglucoside	0·25	0·35	
3,5-diglucoside	0·31	0·23	
3-rhamnoglucoside 5-glucoside	0·29	0·40	
CYANIDIN			
3-monoglucoside	0·38	0·07	red
3-rhamnoglucoside	0·37	0·19	
3-xyloglucoside	0·36	0·24	
3-diglucoside	0·33	0·34	
3,5-diglucoside	0·28	0·16	
3-rhamnoglucoside 5-glucoside	0·25	0·36	
3-xyloglucoside 5-glucoside	0·19	0·41	
PEONIDIN			
3-monoglucoside	0·41	0·09	pink
3,5-diglucoside	0·31	0·17	
3-rhamnoglucoside 5-glucoside	0·29	0·37	
DELPHINIDIN			
3-monoglucoside	0·26	0·03	purple
3-rhamnoglucoside	0·30	0·11	
3,5-diglucoside	0·15	0·08	
PETUNIDIN			
3-monoglucoside	0·35	0·04	purple
3-monorhamnoside	0·40	0·10	
3-rhamnoglucoside	0·35	0·13	
3,5-diglucoside	0·24	0·08	
3-rhamnoglucoside 5-glucoside	0·23	0·37	
MALVIDIN			
3-monoglucoside	0·38	0·06	mauve
3-monorhamnoside	0·39	0·11	
3,5-diglucoside	0·31	0·13	
3-rhamnoside 5-glucoside	0·31	0·34	
3-rhamnoglucoside 5-glucoside	0·30	0·40	

in the rates of hydrolysis of different glycosidic links in anthocyanins as there are with the flavones and flavonols (5.10); as indicated in Table 17, 30 to 40 minutes are sufficient for achieving complete hydrolysis.

The aglycone is extracted with just sufficient amyl alcohol to form an upper layer above the aqueous hydrolysate. Identification is carried out by chromatography in Forestal, butanol–acetic acid and butanol–hydrochloric acid (Table 26) and by visible spectrophotometry (6.6) after diluting the amyl alcohol layer with 95% ethanol. The reaction with aluminium chloride can also be used (6.5). Visible spectroscopy of the six principal anthocyanidins indicates the number of hydroxyl substituents in the B ring (Table 23) and a blue colour with $AlCl_3$ indicates the presence of an *ortho*-dihydroxylic B ring. Identification can finally be confirmed by alkaline fusion, followed by characterisation of the benzoic acid derived from the B ring (6.5).

6.15. *Glycoside identification*

Sugar identification Qualitative analysis of the sugars of anthocyanins can be carried out by the same method as described for flavones (5.9). If solvents containing mineral acid are used during the purification, a sugar arabinose is formed as an artifact from the chromatography paper. By making suitable measurements, the sugar: aglycone molecular ratios can similarly be determined (5.9).

Controlled hydrolysis of the glycosides During controlled hydrolysis of a glycoside under appropriate conditions the sugar molecules are liberated one after another and, before the appearance of the aglycone, different intermediate glycosides are formed, the number and nature of which allows one to identify the original glycoside. This technique, which was developed by Abe and Hayashi (1956) and by Harborne (1959a), is much more successful with anthocyanins than with flavones and their derivatives. The chromatogram illustrated in Fig. 18, taken from the work of Harborne, shows, after controlled hydrolysis, the formation of three pigments from a 3,5-diglucoside, of four pigments from a 3-triglucoside and finally of six pigments from a 3-diglucoside-5-glucoside. Under the same conditions, a 3-monoglucoside forms the aglycone directly.

The following procedure for partial hydrolysis has been used by the author (P. Ribéreau-Gayon, 1959): after elution, the glycoside under study is obtained in 15% acetic acid solution (3.14). The solvent is removed *in vacuo*, the dry residue is taken up in one volume (e.g. 0·2 ml) of 1% HCl and two volumes of methanol and one volume of conc. HCl are then added. This is heated on a water bath at 70° and samples are taken every 30 minutes during two hours. The samples are chromatographed in the solvent recommended for this purpose by Abe and Hayashi (1956), i.e. the upper layer of *n*-butanol–conc. HCl–water (7 : 2 : 5). By taking samples at appropriate time intervals, it is possible to obtain all the intermediate products of hydrolysis.

Original Glucoside	3,5-di-glucoside		3-di-glucoside	3-mono-glucoside	5-mono-glucoside	aglycone
3-diglucoside			original glucoside	product of hydrolysis		product of hydrolysis
3-triglucoside	original glucoside		product of hydrolysis	product of hydrolysis		product of hydrolysis
3,5-diglucoside		original glucoside		product of hydrolysis	product of hydrolysis	product of hydrolysis
3-diglucoside-5-glucoside	original glucoside	product of hydrolysis	product of hydrolysis	product of hydrolysis	product of hydrolysis	product of hydrolysis
			BUTANOL-ACETIC →			
R_f values	0·25	0·31	0·36	0·44	0·51	0·72

(hatched spot) = original glucoside (open spot) = product of hydrolysis

Fig. 18. *A chromatogram of the products of controlled acid hydrolysis of some pelargonidin glucosides* (*Harborne, 1959a*)

Enzymic hydrolysis of the glycosides (Harborne, 1962b) Unlike other glucosides, anthocyanins are not hydrolysed by β-glucosidase. They are however hydrolysed by specific enzymes called anthocyanases, probably widespread in nature, two of which have been characterised: one from a fungal source (an extract from *Aspergillus niger*; Huang, 1955) and the other from a plant source (an extract from fresh cocoa beans; Forsyth and Quesnel, 1957a). These enzyme preparations can be used in identifying anthocyanins since they have different specificities.

The plant enzyme hydrolyses galactosides and arabinosides more quickly than glucosides and xylosides. The fungal enzyme does not distinguish between these four substrates. It attacks anthocyanins more rapidly, however, if they have a simple structure; 3-monoglucosides are hydrolysed in one to two hours while acylated anthocyanins are unchanged after 48 hours. Also, anthocyanins containing rhamnose are especially resistant to attack; by contrast, rhamnosides are much more susceptible to acid hydrolysis than 3-glucosides.

The reaction is carried out in acid buffer pH 3·95; under these conditions the anthocyanidins liberated are unstable (6.4) and there is decolourisation. Anthocyanins can also be oxidised and decolorised by phenolase enzymes (2.6); however, malvidin which lacks a free *o*-dihydroxyl is not affected.

Location of the sugar attachments Hydrogen peroxide oxidation provides a method of identifying the sugar molecules attached in the 3-position (6.5). In addition, consideration of the absorption spectrum can be used to confirm the results of controlled hydrolysis for it shows whether or not there is substitution in the 5-position. The ratio of the absorbances at 440 nm and at the visible maximum is doubled when the 5-hydroxyl is free as compared to when there is a 5-sugar present (6.6).

6.16. *Identification of acylated glycosides*

Chromatographic characteristics It is necessary to point out at once that acylated anthocyanins are more difficult to separate by paper chromatography than the simple glycosides. When several acylated pigments are present in the same plant extract, it can be difficult, occasionally impossible, to separate the individual compounds.

In order to characterise an acylated anthocyanin, chromatography in butanol–acetic acid is effected before and after alkaline hydrolysis; in this solvent, simple glycosides have lower R_f values than acylated glycosides. If alkaline hydrolysis has no effect on R_f, then the pigment is not acylated.

The acyl link is more labile than the glycosidic link and partial hydrolysis occurs during purification. Two spots appear on chromatograms, the major one being the acylated pigment and the other the corresponding simple glycoside. In fact, it is impossible to purify these pigments without some hydrolysis occurring.

The chromatography of complex anthocyanins is best carried out in butanol-based solvents (butanol–acetic acid, butanol–hydrochloric acid) in which R_f's are higher than the corresponding simple glycosides (Table 28). By contrast, in aqueous solvents, the R_f's are of the same order.

Identification of the acyl residues The same techniques are used as for the identification of the ester groups in cinnamic acids (3.5). Alkaline hydrolysis is carried out at room temperature in 2N NaOH for 4 hours in an atmosphere of nitrogen. After acidification, the cinnamic acids are extracted into ether and then identified by chromatography (4.5, 4.6). If it is wished to recover the simple glycoside liberated, acidification should be carried out with cationic ion exchange resin (Dowex 50).

Since anthocyanins are easily oxidised in alkaline solution, the yield is usually low.

Nature of the acyl link Harborne (1964a) has shown that the acyl group (*p*-coumaric, caffeic or ferulic acids) is generally esterified with a hydroxyl group of the sugar attached in the 3-position. When this sugar is a disaccharide, an acyl group can be attached to each of the monosaccharide components. Acylated 3-monoglycosides are relatively rare; the most common types are acylated 3,5-diglucosides and 3-diglycoside-5-monoglucosides.

Table 28. *R_f values of acylated anthocyanins* (Harborne, 1959a)

Acylated glycosides					
Aglycone	Sugars*	Acyl group	Butanol-acetic	BuHCl	1% HCl
pelargonidin	3RG5G	*p*-coumaric	0·37	0·43	0·27
cyanidin	3RG5G	*p*-coumaric	0·32	0·26	0·22
peonidin	3RG5G	*p*-coumaric	0·34	0·31	0·22
delphinidin	3RG5G	*p*-coumaric	0·31	0·24	0·19
petunidin	3RG5G	*p*-coumaric	0·32	0·26	0·19
malvidin	3RG5G	*p*-coumaric	0·36	0·28	0·20
pelargonidin	3G5G	*p*-coumaric	0·40	0·46	0·19
cyanidin	3G5G	*p*-coumaric	0·35	0·34	0·11
delphinidin	3G5G	*p*-coumaric	0·30	0·22	0·05
pelargonidin	3G5G	caffeic	0·37	0·37	0·17
pelargonidin	3GG5G	*p*-coumaric	0·34	0·34	0·49
pelargonidin	3GG5G	ferulic	0·34	0·26	0·49

* Abbreviations: 3RG5G = 3-rhamnoglucosido-5-glucoside; 3G5G = 3,5-diglucoside; 3GG5G = 3-diglucosido-5-glucoside.

Determination of the acyl linkage is carried out by hydrogen peroxide oxidation (6.5), which liberates only the sugar in the 3-position. Acylated sugars are obtained and these can be identified by comparison with reference compounds; one can also identify the products of their alkaline hydrolysis using the methods described for cinnamic acid esters (4.7).

In addition, Albach *et al.* (1965) have isolated malvidin 3-glucoside acylated with *p*-coumaric acid from grapes and shown that the acid is linked to the 4-position of the sugar. In order to do this, the authors methylated the acylated pigment and on hydrolysis obtained a partially methylated sugar with a free hydroxyl at the position of the linkage. This sugar was identified by gas chromatographic comparison with synthetic markers.

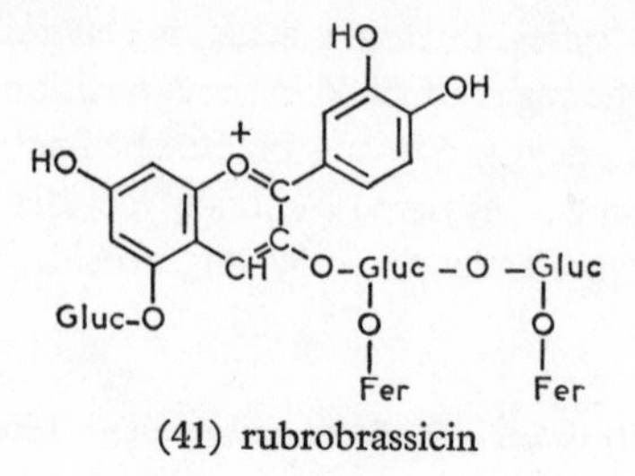

(41) rubrobrassicin

Use of spectroscopy It has already been mentioned (6.6) that the ratio of the maximal cinnamic acid absorption at 300 to 330 nm to that at the anthocyanin maximum at 500 to 530 nm gives a precise measurement of the number of acyl groups present in an acylated pigment. This method has been used by Harborne (1964a) to show that rubrobrassicin (41), the anthocyanin of red cabbage, contains three glucose and two ferulic acid residues linked to the molecule of cyanidin.

7
The Tannins

Introduction

It is difficult to define the term 'tannin' as it is used in plant chemistry in a concise yet rigorous way. The tannins are generally taken to include a whole group of substances which have certain physical and chemical properties in common but which are certainly not related structurally.

Seguin introduced the term 'tannin' in 1796 (White, 1957) to describe the chemical constituents in galls which were responsible for transforming fresh hide into impermeable non-rotting leather. The manufacture of leather had been known since time immemorial but had been attributed to some physical change rather than the action of some defined chemical substances.

Later, it was shown that the natural tanning agents used by the leather industry possessed phenolic properties (e.g. oxidation by cold permanganate, production of colours with iron salts). Constituents having such properties were subsequently identified from a large number of plants and, by analogy, were called tannins or tannoids (analogous to tannins) although their ability to leather hides was never tested. Thus the definition of tannins has been enlarged to cover a whole mass of constituents which merely give general phenolic reactions such as black or green colours with iron salts. Using this broad definition, some authors have been led to include as tannins substances like chlorogenic acid and catechin merely on the basis of their having phenolic properties, in spite of the fact that they do not have the ability to tan hides. It is clear that such a definition of tannins is too wide and should not be retained.

It would not be very useful, however, either for botanists or for agricultural scientists, to define as tannins only those substances which are capable of tanning hides. In order to reach a more satisfactory definition, therefore, it is necessary to consider which chemical and physical characters are essential to give tannins their necessary properties.

The essential property of tannins is their ability to combine with proteins and other polymers such as cellulose or pectin. In the case of tanning itself, the collagen chains in the hide are cross-linked by the tannins to give leather. This is also important in other properties of tannins. For example, the inhibition of enzymes is caused by the tannin combining with the protein moiety; in astringency the 'precipitation' of the proteins and glycoproteins of the saliva by tannins, reduces its lubricant properties (Bate-Smith, 1954d; Joslyn and Goldstein, 1964b). Using this property, tannins can be removed from solution with insoluble polymers such as powdered polyamide (nylon 66 or nylon 6; Harris and Ricketts, 1959), keratin (Vancraenenbroeck and Lontie, 1963) or polyvinyl-pyrrolidone (PVP, Polyclar AT; McFarlane *et al.*, 1955; Chapon *et al.*, 1961; Andersen and Sowers, 1968).

It has been suggested (Swain, 1965b; Loomis and Bataille, 1966) that the formation of the tannin-protein or other polymer combinations involve three types of bonding:

(*a*) Hydrogen bonds (2.3) between the phenolic group of the tannins and receptor groups (—NH—, —CO—, or —OH) of the protein or other polymer.

(*b*) Ionic bonds between anionic groups of the tannins (ionised phenolic or carboxylic groups) and cationic groups (e.g. ϵ-amino group of lysine) of the proteins.

(*c*) Covalent links formed by the reaction between quinones, which either form part of the tannin structure (7.8) or are produced by oxidation (2.6), and some reactive group in the protein or other polymer. These latter bonds are highly important since they impart great stability to the tannin-protein complex.

If the complex is to be sufficiently stable, the tannin must have enough phenolic groups to form links with a single protein chain or between two or more protein chains. On the other hand, the tannin must not be too large or it will be unable to orient itself closely enough to the protein to give stable linkages a chance to form. It is considered that the most stable tannin protein complexes are those formed by tannins with a molecular weight of between 500 and 2,000.

In theory, compounds within this molecular weight range are smaller than colloids. However, the presence of numerous hydroxyl groups in the tannins leads to formation of hydrogen-bonded aggregates of tannin molecules of near colloidal size. The individual molecules of tannins will disaggregate in the presence of proteins to form more stable complexes. The formation of such aggregates explains, in part at least, why it is so difficult to fractionate the individual substances present in tannin extracts.

One is thus led to seek a definition of the term tannin which will be useful in phytochemistry and which takes into account the characteristic properties of the substances rather than merely referring to the leather industry. One such definition has been given by Swain and Bate-Smith (1962):

> 'It seems more realistic to define as tannins all naturally occurring substances which have chemical and physical properties akin to those which are capable of making leather. This means that they would be water-soluble phenolic compounds, have molecular weights lying between 500 and 3,000, and, besides giving the usual phenolic reactions have special properties such as the ability to precipitate alkaloids, gelatin and other proteins.'

The authors added:

> 'It is obvious that such a definition will include molecules which are not tannins in the commercial sense of being economically important in the tanning of hides, but will exclude a large number of substances whose only relationship to the tannins is their capacity to reduce alkaline permanganate or give colours with ferric salts.'

All chemical classifications of tannins are necessarily arbitrary. However, it is usual to distinguish between 'hydrolysable tannins' and 'condensed tannins' which correspond to two well differentiated structural types. Hydrolysable tannins consist of a carbohydrate core, the hydroxyl groups of which are esterified by gallic acid or one of its derivatives (hence the name gallotannin which is sometimes given to the group, see below). As their name suggests, they are readily hydrolysed either chemically or enzymically. Condensed tannins are formed by the condensation of hydroxyflavans (catechins or leucoanthocyanins); they are often also called 'catechin tannins' and are resistant to hydrolysis.

According to White (1957), world production of commercial tannins is of the order of 500,000 tons per year of which four-fifths are condensed tannins. The principle sources are extracts of quebracho wood

(*Shinopsis lorentzii*) and mimosa bark (*Acacia mimosa*) for condensed tannins, and chestnut wood (*Castanea sativa* and *C. dentata*) and dry myrobalan fruits (*Terminalia chebula*) for the hydrolysable tannins.

Industrially, tannins are used in several different ways other than in a tannery (Hathway, 1962). Most of these do not depend on their property of combining with proteins. For example, they are used to protect fishing nets, to reduce the viscosity of the mud in oil-well drilling (this accounts for 40% of the tannins used in the United States); and to protect metal drainage pipes. Finally, condensed tannins are used as one of the main starting points for the production of phenol-formaldehyde resins (2.9).

Hydrolysable tannins*

7.1. *Introduction*

Hydrolysable tannins are esters of sugars and phenolic acids or their derivatives; the sugar is usually glucose, but in some cases polysaccharides have been identified. Acidic, basic or enzymatic hydrolysis is relatively easy and partial hydrolysis often occurs spontaneously during extraction and purification.

These tannins are usually divided into 'gallotannins', which on hydrolysis yield gallic acid (1) as the only phenolic moiety, and the more numerous 'ellagitannins', which under the same conditions give besides gallic acid one or more of its derivatives, of which ellagic acid (2)

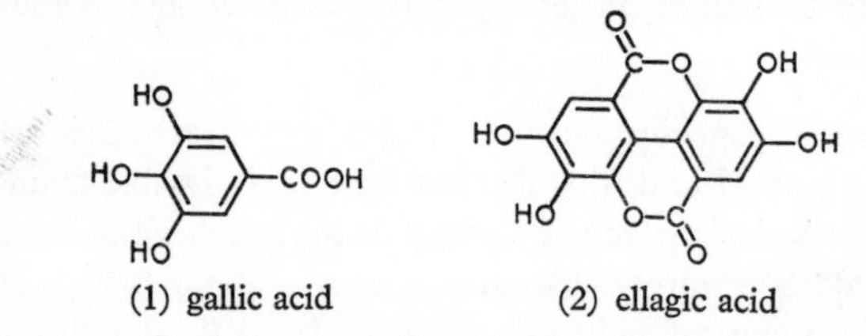

(1) gallic acid (2) ellagic acid

is the most important. These tannins appear to be somewhat generally distributed in the plant kingdom although detailed studies have only been made on species which give rise to commercial tannins. However, they are much less common than the condensed tannins.

Studies on the chemistry of the hydrolysable tannins have been greatly aided by the use of paper chromatography. For example, it has been shown that the majority of previously isolated components which

* This account of the hydrolysable tannins owes much to articles by White (1956, 1957, 1958), Jurd (1962b) and Haslam (1966, 1967).

had at the time been considered pure were really mixtures, readily resolvable by two-dimensional chromatography (Fig. 19). Certain individual substances in these mixtures have now been isolated in a pure state and their structures determined. It should be noted that the condensed tannins give streaks rather than the discrete spots found on chromatograms of hydrolysable tannins (Fig. 19).

Most investigators have used 5-10% acetic acid and butanol-acetic (5.5) as solvents in two-dimensional paper chromatography and revealed the tannins with ammoniacal silver nitrate (2.6) or ferric chloride–potassium ferricyanide (3.10).

7.2. *The gallotannins*

The gallotannins are esters of glucose, or a polysaccharide, and gallic acid or *m*-digallic acid (3) which itself gives gallic acid on hydrolysis. The exact structures of tannins of this type have not been elucidated. However, several simple galloylglucoses have been studied, obtained either directly from natural extracts or by partial degradation of more complex molecules. The chief of these are 1-galloyl-, 3,6-digalloyl- and 1,3,6-trigalloylglucose (4).

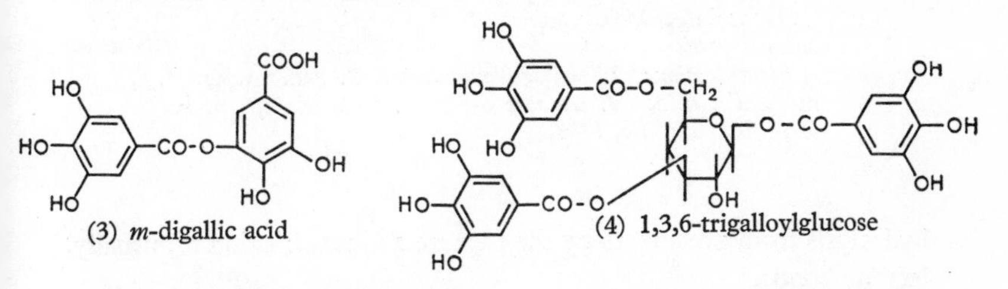

(3) *m*-digallic acid

(4) 1,3,6-trigalloylglucose

The chief commercial tannin of this type is tannic acid from plant galls. Paper chromatography (White, 1958) has shown that it is impure (Fig. 19). It contains 21% gallic acid, 7% *m*-digallic acid, 3% pentagalloylglucose, 1% trigallic acid and 70% of the main gallotannin. The structure of the latter has not been elucidated but it contains from 8 to 10 moles of gallic acid per mole of glucose.

7.3. *The ellagitannins*

This class of hydrolysable tannins is much better known than that described above. They are characterised by the fact that on acid hydrolysis they yield, besides gallic acid itself, several derivatives of the latter compound. These congeners are not necessarily combined directly with glucose in the original tannin, but are formed after

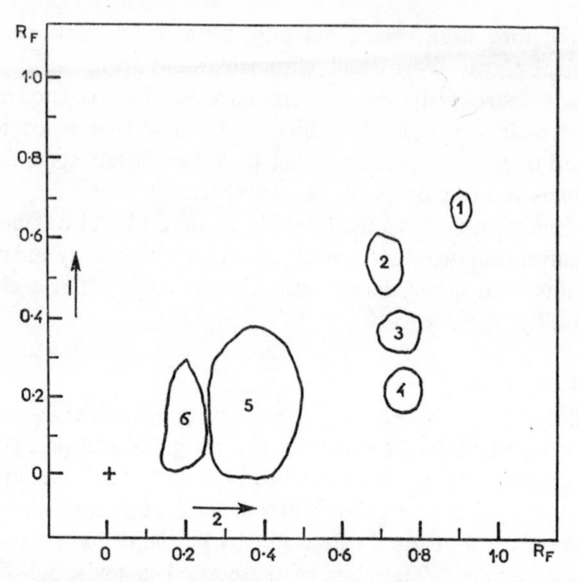

Fig. 19. *Two-dimensional chromatogram of tannic acid (hydrolysable tannin from galls on* Rhus semialta *and* Quercus infectoria). *Solvent 1, 10% acetic acid; solvent 2, butanol-acetic*

Identification: 1. pentagalloylglucose; 2. gallic acid; 3. m-*digallic acid; 4. trigallic acid; 5. gallotannin; 6. not identified. (White, 1958)*

hydrolysis from precursors by the rupture and reformation of, usually, lactone bonds.

The formulae and structural relationships of the different compounds which are obtained by hydrolysing various different ellagitannins are shown in Fig. 20. The most important compound is ellagic acid (2) formed by lactonisation of hexahydroxydiphenic acid (5).

The best known ellagitannins are chebulagic acid (6) and chebulinic acid (7). The first gives equimolar proportions of glucose (ring *a*), gallic acid (*b*), ellagic acid (*c* and *d*), and chebulic acid (*e* and *f*) on acid hydrolysis. The second tannin has a closely related structure giving two more moles of gallic acid instead of the one of ellagic acid due to the fact that the aromatic rings *c* and *d* (7) are not linked.

Ellagic acid can be readily identified in hydrolysed plant extracts by paper chromatography using Forestal solvent (3.6) (Bate-Smith, 1956a).

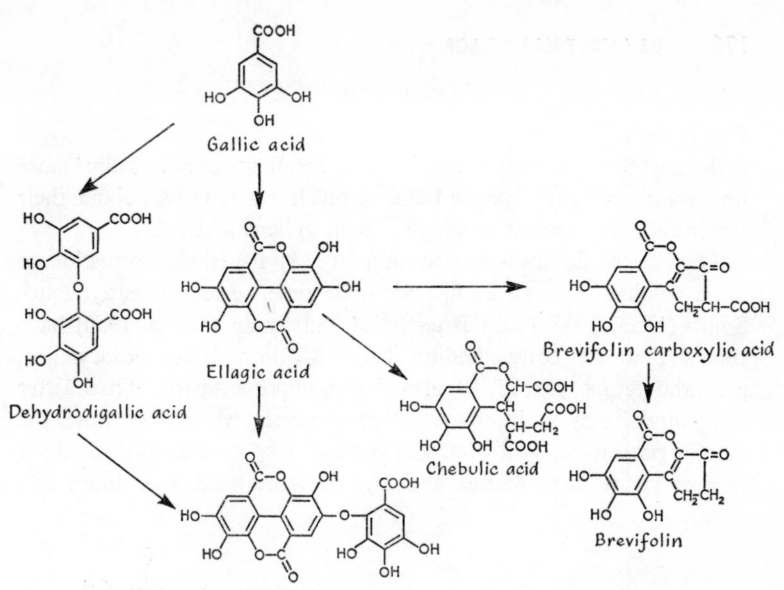

Fig. 20. *Formulae of various phenolic compounds obtained on hydrolysis of ellagitannins.*

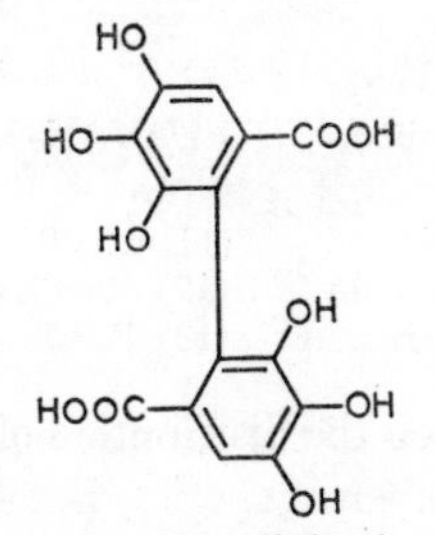

(5) hexahydroxydiphenic acid

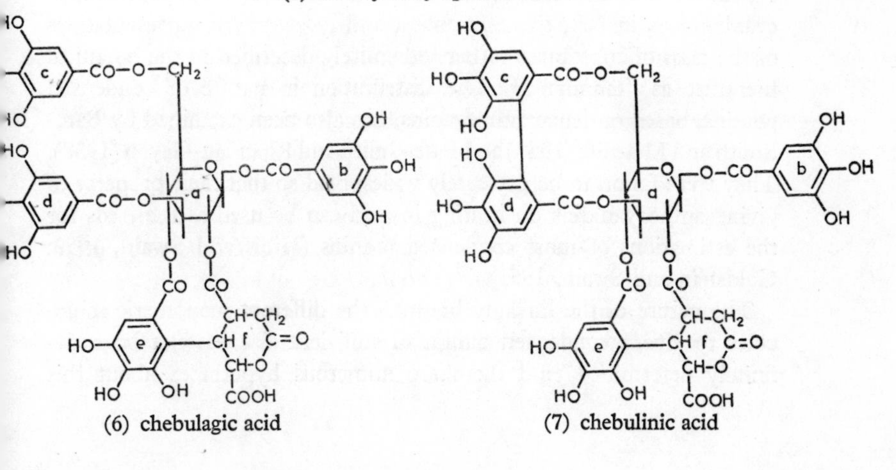

Condensed tannins

7.4. *Introduction*

Although the condensed tannins are certainly commercially more important than hydrolysable tannins, much less is known about their structure, many aspects of which remain to be elucidated.

The role of flavan-3-ols (catechins) (8, 9, 10) in the formation of these tannins has been known for some time. More recently, Bate-Smith (1953, 1954b) and Bate-Smith and Swain (1954b, 1956) have shown that the corresponding flavan-3,4-diols (leucoanthocyanins, proanthocyanins) (11, 12, 13) also play an important part. These latter compounds, and their condensation products which constitute the tannins proper, are easily detected because they are transformed, albeit in poor yield, into coloured anthocyanidins by heating in dilute acid solution.

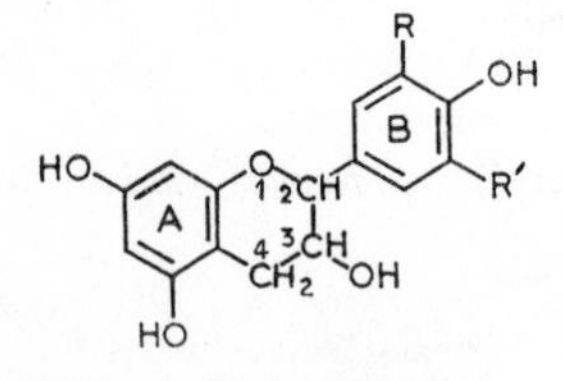

FLAVAN-3-OLS (catechins):

(8) R=R′=H; afzelechin
(9) R=OH, R′=H; catechin
(10) R=R′=OH; gallocatechin

FLAVAN-3,4-DIOLS (leucoanthocyanidins):

(11) R=R′=H; leucopelargonidin
(12) R=OH, R′=H; leucocyanidin
(13) R=R′=OH; leucodelphinidin

At the end of a study on the distribution of the leucoanthocyanins in the leaves of a large number of species, Bate-Smith and Lerner (1954) stated: 'It is difficult to avoid the conclusion that they (leucoanthocyanins) are, in fact, the commonest and most typical representatives of the class of substances rather indefinitely described in the botanical literature as "tannins" '. The distribution in nature of condensed tannins, based on leucoanthocyanins, has also been examined by Bate-Smith and Metcalfe (1957) and Bate-Smith and Ribéreau-Gayon (1958). They were found to be extremely widespread so that their property of giving anthocyanidins on heating in acid can be used as the basis for the estimations of most condensed tannins (Hillis and Swain, 1959; Goldstein and Swain, 1963).

The nature of the linkages between the different monomeric molecules to give a condensed tannin of sufficient size has not been definitely determined, and there are numerous hypotheses about this problem Every plant so far examined contains several different tannin-like compounds which vary either in the kind or in the mode of polymerisation of the elementary monomers which are involved in the structure. The problem of separating such different tannins which are found in plant extracts has not been solved; paper chromatography (with butanol–acetic) gives diffuse streaks probably due to the formation of complexes between the tannins and the cellulose. The only compounds which can be satisfactorily separated by paper chromatography are the monomeric flavans and some dimers (biflavans, 7.7) which do not have any tannin-like properties. It is true that so far no condensed tannin has yet been isolated in a pure state.

In the following sections the flavans which are known to be involved in condensed tannin formation will be first described. The flavanonols (14, 15) and hydroxystilbenes (16) which may also be incorporated

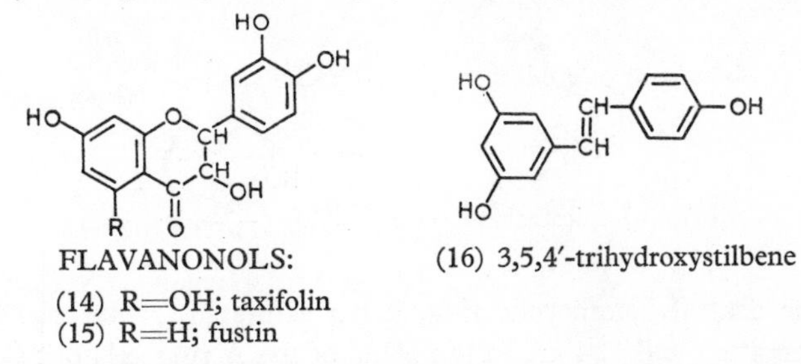

FLAVANONOLS:
(14) R=OH; taxifolin
(15) R=H; fustin

(16) 3,5,4′-trihydroxystilbene

(Hathway, 1962) will not be dealt with. The different hypotheses which have been put forward to explain the mode of polymerisation of the monomers and thus the properties of the tannin will be discussed next. Finally, in the last section (7.9), the methods used for studying tannins in plant tissues are outlined. The emphasis will be on the general interest of tannins in plant biochemistry, especially their involvement in food quality, and not on their use in leathering hides.

7.5. *The flavan-3-ols (catechins)*

Catechins have the structure of flavan-3-ols and are related to anthocyanins and flavones. However, catechins do not generally exist in nature as glycosides. The most common members of the group (8, 9, 10) differ only in the number of hydroxyl groups (1, 2 or 3) in the phenyl B ring and these groups are never methylated. The term 'catechin' refers specifically to the flavan-3-ol which has two hydroxyl groups in the side ring (9). All the compounds have two asymmetric carbon atoms (C-2 and C-3) thus giving four optical isomers; in the case of the catechin series these configurations are (+)-catechin (17), (−)-catechin (18), (+)-epicatechin (19) and (−)-epicatechin (20) (Whalley, 1962). (+)-Catechin and (−)-epicatechin are the common naturally occurring forms.

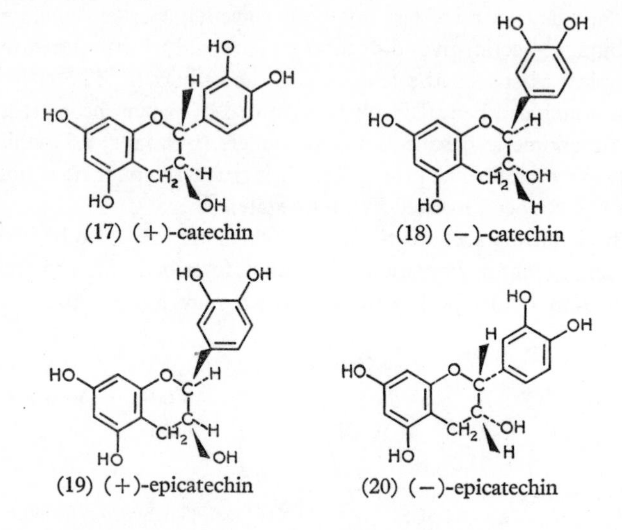

(17) (+)-catechin

(18) (−)-catechin

(19) (+)-epicatechin

(20) (−)-epicatechin

The central heterocyclic ring, being saturated, is not planar; the oxygen atom and C-4 are in the plane of the A ring (21) but C-2 and C-3 lie either above or below the ring (21, 22). Several considerations, notably the IR spectra, show that there is strong hydrogen bonding between the hydroxyl group at C-3 and the heterocyclic oxygen (Hathway, 1962) in both (+)-catechin and (−)-epicatechin, and the configuration of these isomers is thus represented by (21) and (22). The configuration at C-2 is the same in each case and this appears to be the same in related naturally-occurring compounds. The other isomers (epimeric at C-2) can be obtained by heating the natural forms in aqueous solution.

The catechins also include gallic acid esters (23, 24) with the acid moiety attached to the OH group at C-3. The liberation of gallic acid on hydrolysis is thus not confined to the hydrolysable tannins.

The main difference between this class of compounds and the flavan-3,4-diols (7.6) is the fact that on heating in acid solution the catechins give yellow-brown insoluble products of high molecular weight, whereas the flavan-3,4-diols give, besides these phlobaphenes, some anthocyanidin.

If one wishes to examine the catechins present in natural extracts, it is essential to avoid acid hydrolysis. The catechins can be extracted with ether or ethyl acetate and separated, after concentration, by two-dimensional chromatography in butanol–acetic and dilute aqueous

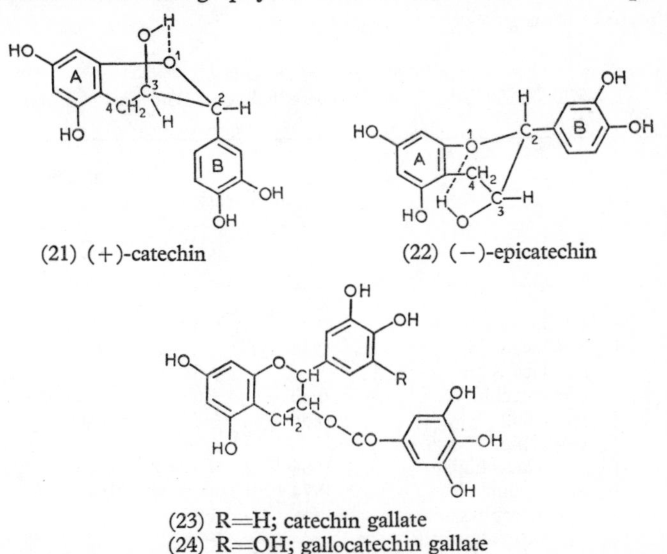

(21) (+)-catechin

(22) (−)-epicatechin

(23) R═H; catechin gallate
(24) R═OH; gallocatechin gallate

acetic acid (2-15%) (5.3). The compounds can be revealed by spraying with a solution of 10% vanillin in concentrated hydrochloric acid which gives rise to red coloured spots (2.9). The R_f values of several flavan-3-ols is given in Table 29. The variations obtained by the different authors should be noted.

7.6. *Flavan-3,4-diols (leucoanthocyanidins)*

The term 'leucoanthocyanin' is particularly difficult to define. Much of the confusion is due to the usage of the term in an imprecise manner.

The terms 'leucoanthocyanidin' and 'leucoanthocyanin' have been used interchangeably over the last twenty years or so for those natural products which yield anthocyanidins on heating in acid solution. The second term suggests that the compounds may be glycosidic (cf. anthocyanin) but this has not been confirmed.

According to the literature, the leucoanthocyanins were first discovered by Rosenheim in 1920. However, ten years earlier, Laborde (1910) (then Deputy-Director of the Agricultural and Viticultural Institute of Bordeaux), in a little known but remarkably accurate paper, demonstrated the presence of colourless compounds in wine and in various parts of the grape vine which gave coloured anthocyanins on heating in acid solution. He also established a relationship between these bodies and tannins and used the colour-forming reaction to estimate tannins in wine.

Table 29. *R_f's of the catechins*
(I.—Hillis and Carle, 1960; II.—Bradfield and Bate-Smith, 1950)

	Butanol-acetic (I)	Butanol-acetic (II)	Acetic acid 6% (I)
(−)-epigallocatechin	0·24	0·47	0·30
(+)-gallocatechin	0·34		0·40
(±)-gallocatechin		0·57	
(−)-epicatechin	0·43	0·65	0·29
(+)-catechin	0·56	0·76	0·40
(±)-catechin		0·74	
(−)-epiafzelechin	0·64		0·43
(+)-afzelechin	0·74		0·47
(−)-epicatechin gallate		0·86	
(−)-gallocatechin gallate		0·72	

For a number of years it has been recognised that these compounds play an important role in the structure of tannins. It is known that the leucoanthocyanin polymers, which constitute the tannins, give anthocyanidins on heating in dilute acid almost as easily as do the monomers. Consequently the expressions 'leucoanthocyanin' and 'leucoanthocyanidin', defined as above, have lead to some confusion since the compounds included are not all tannins. On the other hand, most experts recognise that the expressions are imprecise, and that the facile transformation to anthocyanidins is not their most important property. For example, Bate-Smith and Swain (1953) stated 'it is unfortunate that the colour produced on heating with HCl has hitherto focused attention rather narrowly on the relationship of leucoanthocyanins to the anthocyanidins, a relationship which from the plant physiological point of view, may be quite fortuitous'. It is true, however, that it has been thought for a long time that the compounds were structurally closely related to the anthocyanins. Laborde (1908) and several later authors have put forward the now abandoned hypothesis that the leucoanthocyanins are precursors of anthocyanins. In particular, Laborde believed that this mechanism was responsible for autumnal coloration in leaves.

Swain (1962a) suggested that the polymers formed from flavans should be called 'flavolans' by analogy with the nomenclature of the polysaccharides (araban, dextran), while monomers were given a specific name, for example, leucocyanidin (the compound which gives cyanidin on hot acid treatment). The term 'leucoanthocyanidin' can be used to describe all types of monomer, but the term 'leucoanthocyanin' as a generic expression should be abandoned as it is not specific enough. It has been used for monomers, their condensation products and the constituents of vegetable tannins and does not allow any distinction to be made between them.

Freudenberg and Weinges (1962) proposed the term 'proanthocyanidin' to cover those colourless substances which were transformed into anthocyanidins on heating in acid solution. This expression suffers from the same drawback as the term 'leucoanthocyanin', but has often been used specifically for the biflavans described below (7.7).

Structure and properties of flavan-3,4-diols Swain, in 1954, established that the leucoanthocyanidin monomer had the structure of a flavan-3,4-diol by reducing taxifolin (25) (dihydroquercetin) with sodium borohydride to 5,7,3′,4′-tetrahydroxyflavan-3,4-diol (26), and showing that this was identical in several respects with natural leucocyanidin. The compound was first obtained crystalline by Michaud and Masquelier in 1968. Flavan-3,4-diols can also be synthesised by oxidation of a flavylium salt, not hydroxylated at C-3, by hydrogen

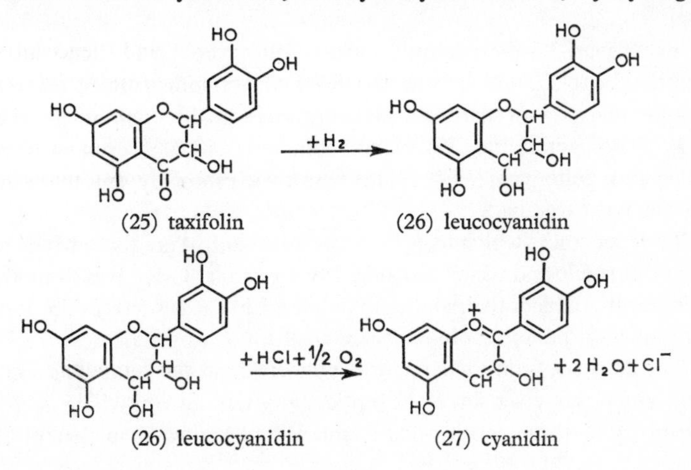

(25) taxifolin

(26) leucocyanidin

(26) leucocyanidin

(27) cyanidin

peroxide (6.5) followed by reduction of the product with sodium borohydride (Jurd, 1968c).

Swain showed that the transformation of leucocyanidin (26) to cyanidin (27) did not take place in an atmosphere of N_2 and concluded oxygen was necessary. The mechanism of the reaction has been studied by Pigman *et al.* (1953), Joslyn and Goldstein (1964a), Monties (1966) and Peri (1967). In every case, the yield of anthocyanidin is quite small; there is a simultaneous condensation of the flavan-3,4-diols to give phlobaphenes and this is the major reaction (80%) leaving, at best, a 20% yield of anthocyanidin.

The condensed tannins (flavolans) are also transformed into anthocyanidins on heating in dilute acid, but the yield is exceedingly small. This reaction distinguishes these substances from the catechins and their polymerisation products which under the same conditions give more highly condensed products without the formation of anthocyanidins (7.5). The absorption spectrum of the products obtained by heating a natural leucocyanidin and catechin in acid solution are shown in Fig. 21 (Swain and Hillis, 1959); the maxima at 550 nm is due to the cyanidin formed, while that at 450 nm is due to phlobaphene. The figure also shows the effect of different solvents on the reaction (7.10).

The different known leucoanthocyanidin monomers have been described by Clark-Lewis (1962). They are all flavan-3,4-diols but it is quite possible that other structural types may exist. For example, Bate-Smith and Swain (1967) demonstrated the presence of a leucoluteolinidin in several grasses. This substance which gave luteolinidin (6.1) on heating in acid was presumed to have the structure of a flavan-4-ol.

The flavan-3,4-diols known include leucocyanidin (12) and leucodelphinidin (13), which have the 5,7-hydroxylation pattern common to the majority of natural flavonoids, teracacidin (28), melacacidin (29), leucofisetinidin (30), leucorobinetinidin (31), cyanomaclurin (32), and peltogynol (33). The first two, especially leucocyanidin, are the most frequent.

The monomers which have been the best studied are those which are the most stable and which lack a hydroxy group at C-5. It is likely that this feature reduces their ability to polymerise (Creasy and Swain, 1965) by reducing the nucleophilic character of the A ring (2.7).

The flavan-3,4-diols have three asymmetric carbon atoms (2, 3 and 4) and thus can exist in eight optically active forms. The stereochemistry of these compounds is much less known than that of the

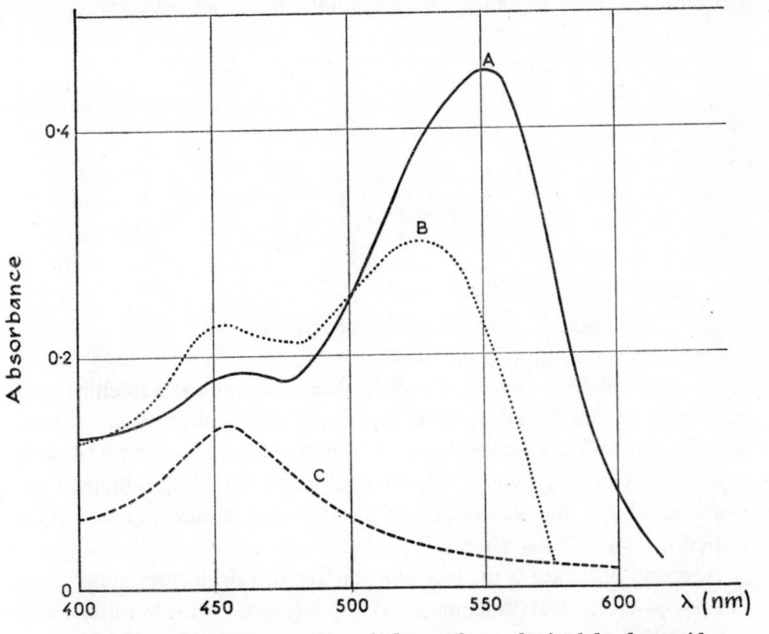

Fig. 21. *Absorption spectra of the products obtained by hot acid treatment of flavans (Swain and Hillis, 1959). A. leucocyanidin in* n-*Butanol-HCl; B. leucocyanidin in acetic acid-HCl; C.* (+)-*catechin in* n-*butanol-HCl*

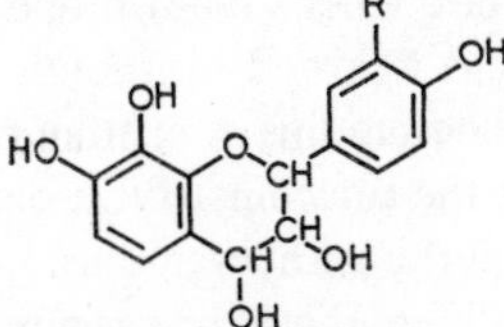

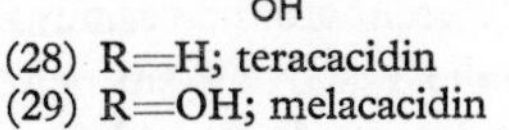

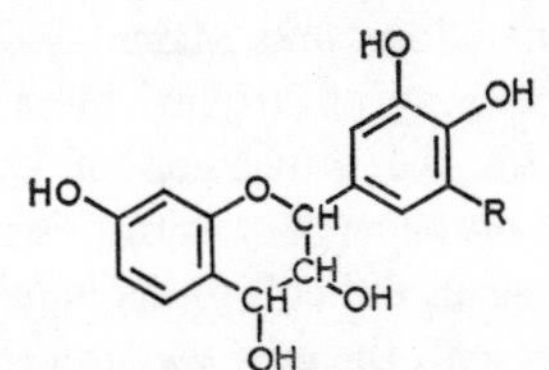

(30) R=H; leucofisetinidin
(31) R=OH; leucorobinetinidin

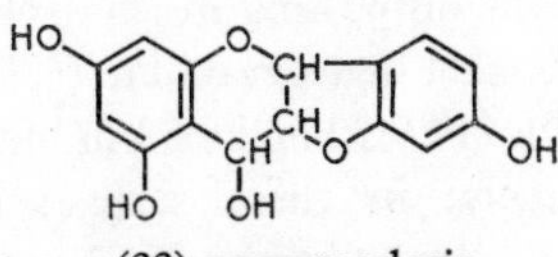

(32) cyanomaclurin

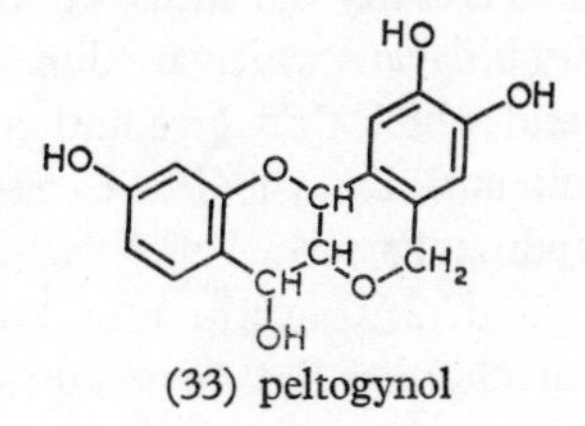

(33) peltogynol

catechins. The absolute configuration of (−)-leucofisetinidin is *trans*2,3-*cis*3,4 (34), and several isomers of teracacidin have been obtained from acacia wood (Drewes and Roux, 1966).

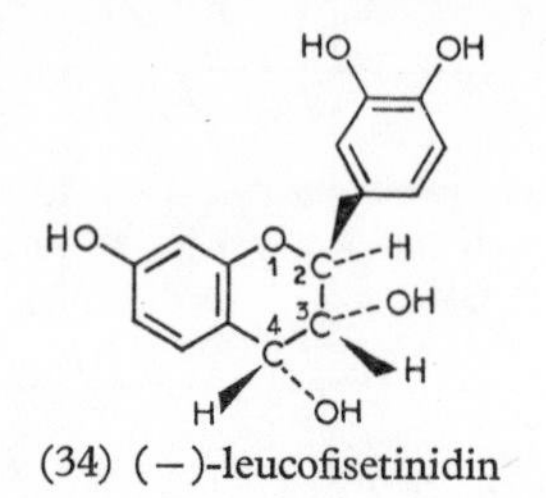

(34) (−)-leucofisetinidin

Chromatography of flavan-3,4-diols The separation and identification of the flavan-3,4-diols on paper chromatograms is more difficult than for the flavan-3-ols (catechins). For one thing, the flavan-3,4-diols readily polymerise, except for those unhydroxylated at C-5, and the isolation of the monomeric forms in sufficient amount is extremely difficult.

A second difficulty is the lack of a suitable discriminatory spray; both flavan-3,4-diols and catechins react equally well with vanillin (7.5). Roux (1958) proposed a spray of 3% *p*-toluenesulphonic acid in ethanol followed by heating at 80-90°C. Under these conditions, leucoanthocyanidins with no hydroxylation at C-5 (e.g. melacacidin or leucofisetinidin) give a very distinctive red colour, but the other leucoanthocyanidins give a brownish reaction little different from that given by the catechins (Roux and Mains, 1960). Masquelier (1964) proposed the use of a spray of 20% trichloracetic acid followed by heating to 90-100°C for the conversion of leucoanthocyanidins to anthocyanidins directly on paper, but with this reagent the catechins give a deep rose colour, again difficult to distinguish from the anthocyanidins.

At present, the only way to properly differentiate the various flavans which react with vanillin is to isolate the reactive spots, elute the components, effect their transformation in a test tube to anthocyanidins and then identify the latter. Using this technique, it is also possible to identify biflavans which are dimers of a flavan-3-ol and a flavan-3,4-diol (7.7) and which thus give anthocyanidins on heating in acid solution.

Roux and Evelyn (1958b) reported that leucocyanidin (5,7,3′,4′-tetrahydroxyflavan-3,4-diol) has R_f values of 0·52 and 0·53 in butanol–acetic and 2% acetic acid respectively; in these same solvents (+)-catechin has R_f 0·76 and 0·39 and (+)-gallocatechin 0·57 and 0·33.

7.7. *The biflavans*

The biflavans, which occur in many fruits, are dimers in which one molecule of a flavan-3-ol is linked to a flavan-3,4-diol, usually leucocyanidin or leucodelphinidin. They were first identified in 1960 by Forsyth and Roberts and have since been extensively investigated (Freudenberg and Weinges, 1961; Geissman and Dittmar, 1965; Creasy and Swain, 1965; Weinges and Freudenberg, 1965; Mayer *et al.*, 1966; Joslyn and Dittmar, 1967; Weinges *et al.*, 1968) often under the name 'proanthocyanidins' proposed by Freudenberg and Weinges (1962) as described in the previous section (7.6). A review of these compounds has been given by Seshadri (1967).

The biflavans separate as well-defined spots on paper or thin layer chromatography, their R_f values in butanol acetic being lower than those of the catechins. Creasy and Swain (1965) studied different biflavans by TLC on cellulose, silica gel and polyamide in several solvents. In 70% methanol on polyamide, monomeric flavans all had R_f of 0·50, and biflavans of 0·63. The two constituent monomers are split by relatively much milder acid treatment than is required for the transformation of flavan-3,4-diols into anthocyanidins. Five minutes in 0·1N HCl at 100°C is usually sufficient.

The biflavans may be intermediates in the formation of condensed tannins, or they may be formed independently by a parallel route. This second hypothesis seems to be the more plausible at present. Polymerisation of flavan-3,4-diols gives polymers of high molecular weight while the reaction between the diols and the corresponding 3-ols yields only dimers (7.8) (Geissman and Dittmar, 1965).

It is certainly true that the biflavans have an important role. On the one hand, the elucidation of their structure has thrown light on that of the condensed tannins themselves, and on the other, the biflavans, having a molecular weight above 500, possess certain tannin-like properties; they are, for example, astringent.

Many structures have been proposed for the biflavans (e.g. 35 and 36), but none has been definitely proven. All that can be said is that the link between the two monomers is labile and readily split by heating in dilute acid. A synthetic biflavan was prepared by Geissman and Yoshimura (1966) which had the structure (36) originally proposed for a natural biflavan by Geissman and Dittmar (1965). This type of structure was confirmed by the condensation of phloroglucinol with flavan-3,4-diols to give compounds such as (37) (Jurd and Lundin, 1968).

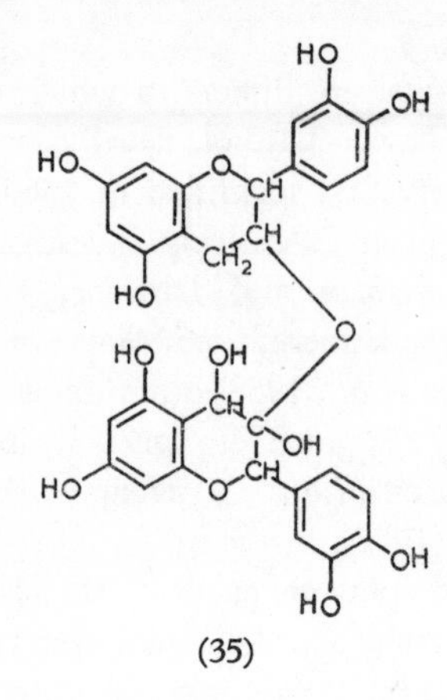

(35)

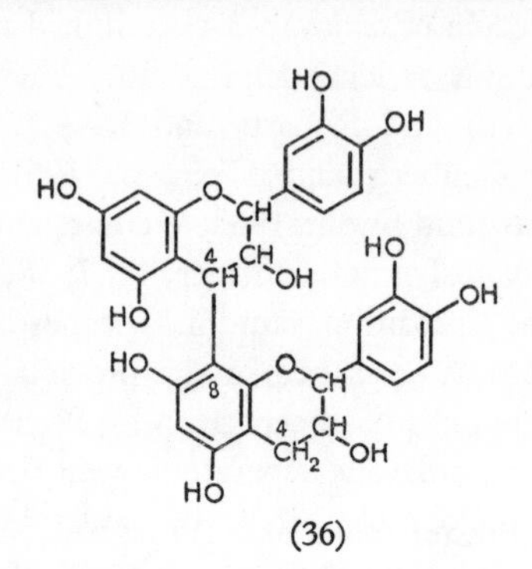

(36)

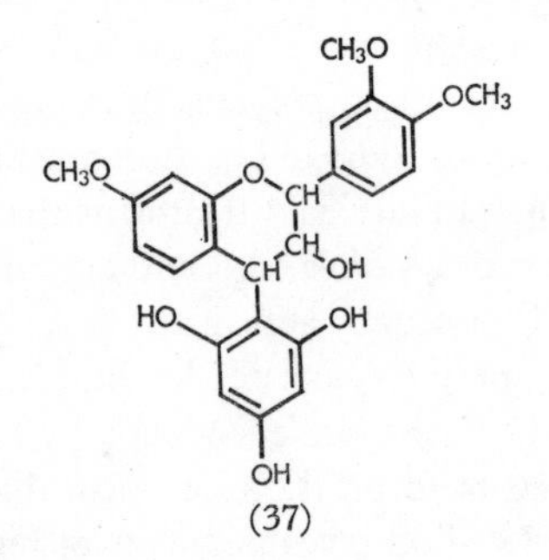

(37)

7.8. *Structure of condensed tannins*

Condensed tannins from plants are a mixture of several polymers which are condensation products of the flavans described above. These polymers have been designated flavolans (7.6) in which the flavan-3,4-diols play a more important role than the flavan-3-ols.

The tannins have a molecular weight of between 500 to 3,000; that of the flavans is of the order of 250 to 300 which means that the flavolans contain 2 to 10 monomers and could thus be called oligomers. Polymerisation plays an important role in determining the properties of the tannins, which vary depending on whether dimers, trimers or tetramers are present. On the other hand, Quesnel (1968) has drawn attention to the importance of the shape of the flavolans which probably plays as great, if not greater, part as the size in determining the properties of these substances. It is thus of particular interest to know the nature of the links in the condensed tannins. Although this problem has been well studied it has not been definitely solved and several different schemes have been put forward.

Freudenberg and Weinges (1962) examined the condensation of flavans in the presence either of cold mineral acid or in hot aqueous solution. Although such conditions of acidity or temperature could not be held to account for the formation of tannins in plant tissues, they might be responsible for the types of change which occur during the processing of foods or beverages such as wine. A knowledge of the reactions involved could also lead to a better understanding of phlobaphene formation in hot acid solution.

In the case of the flavan-3-ols, Freudenberg and Weinges, noted there was an increase in the number of hydroxyl groups on polymerisation presumably due to the addition of water to the O—C(2) bond of the heterocyclic ring thus giving one phenolic and one benzylic alcoholic hydroxyl group extra (38). The latter readily forms a carbonum ion (with a reduced electron density) in acid milieu by the reaction:

$$C_6H_5{-}\overset{|}{\underset{|}{C}}{-}OH + H^+ \rightleftharpoons C_6H_5{-}\overset{|}{\underset{|}{C}}{}^+ + H_2O$$

This ion can then react with a strongly nucleophilic centre (one having an excess of electrons) such as is found at C-6 or C-8 of the flavan A ring of a second molecule of flavan-3-ol (2.7) giving a dimer such as (39)

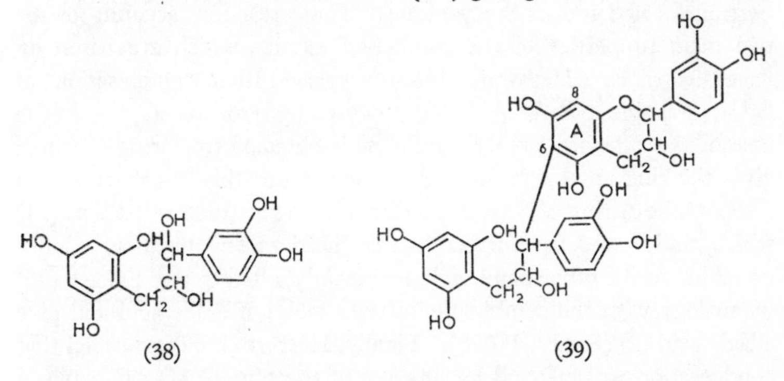

(38) (39)

whose structure was proved by Weinges and Nagel (1968). Such a mechanism could involve a large number of molecules to yield a high molecular weight polymer.

With flavan-3,4-diols, Freudenberg and Weinges (1962) found a much greater ease of polymerisation than for the corresponding flavan-3-ols. They presumed that this was due to the fact that the hydroxyl group at C-4 was also benzylic in nature and could react with the nucleophilic centres of another hydroxyflavan to give a structure like (40). This could then condense with a further flavan molecule through the free OH at C-4 in the lower ring system. The structure (40) is very similar to the biflavan isolated by Geissman and Dittmar (7.7) (36), except the latter cannot polymerise beyond the dimer because the catechin moiety has no hydroxyl group at C-4.

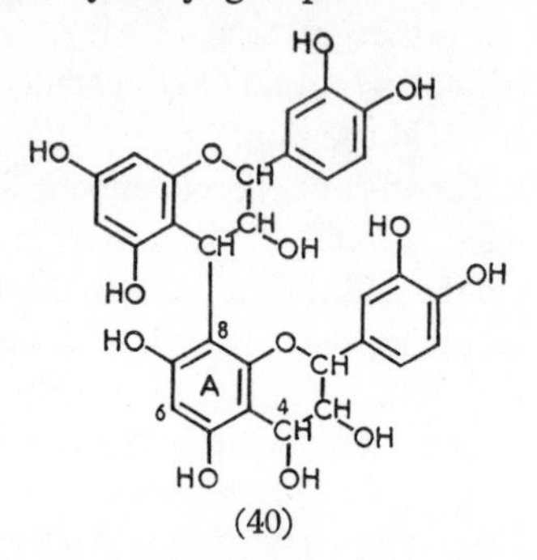

(40)

By treatment with acid, a molecule such as (40) gives carbonium ions and rapidly gives insoluble high molecular weight compounds, the phlobaphenes. But, under the same conditions, certain bonds in the polymer can be broken giving flavan-3,4-diol monomers which can then be transformed into anthocyanidins. These reactions account for the two main properties of the condensed tannins which are based on flavan-3,4-diols. Hathway (1962) suggested that condensation of flavans was due to their oxidation catalysed by tyrosinase (2.6). In the case of the catechin, an *ortho*-quinone is obtained from ring-B which gives the ring an electrophilic character. Thus this can react with a nucleophilic A ring of a second flavanol giving a structure such as (41) which could condense further in this 'head to tail' manner. Gallocatechin, on the other hand, will give mainly a 'tail to tail' linkage (42) by analogy with the combination of two moles of gallic acid and give ellagic acid (2) (Swain, 1965b). Finally, Hergert (1962) proposed that condensation was effected by linkages of the type C—O—C between the hydroxyl group at C-3 or -4 of one flavan molecule and the OH at C-7 of a second (43).

It seems highly likely that, in any one tannin, several different types of linkage exist simultaneously, perhaps involving several different monomers. Creasy and Swain (1965) envisage the possibility of polymerisation in three dimensions. Again, flavans may condense with other types of compound which would thus be incorporated into the tannin. Jurd (1967) showed that catechins could form dimers with certain flavylium salts, especially those lacking an hydroxyl group at C-3. Thus the synthetic compound (44a or 44b) reacts with (+)-catechin (45) in dilute acetic acid to give a colourless product, probably the flavene (48), and an orange dimeric flavlium salt (47), the best yields being obtained from one mole of (45) and two of (44a). It appears

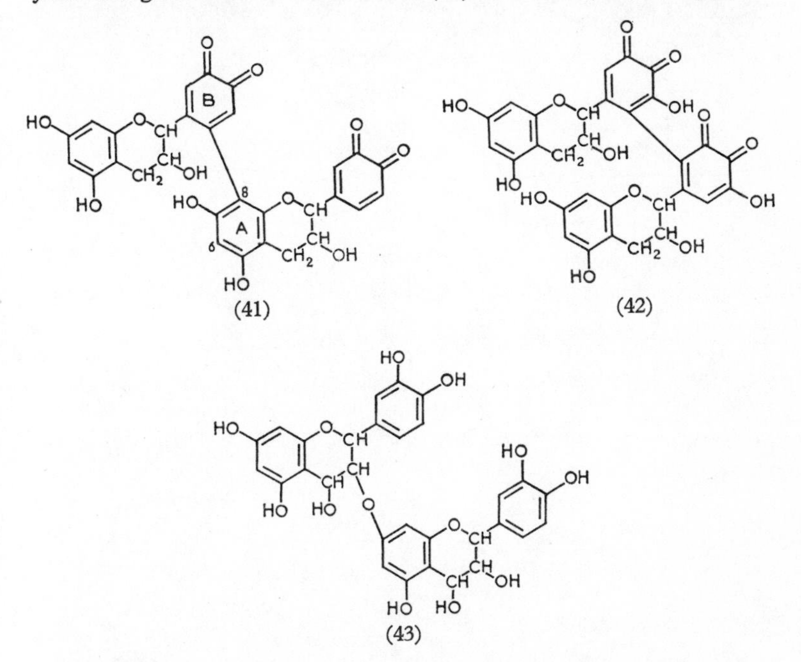

(41) (42) (43)

likely that the dimeric intermediate (46) is first formed which is then oxidised by a second mole of the salt (44a). The flavene (48) is rapidly hydrolysed in aqueous acid solution to give a chalcone (1.1). It is thus probable that, in some plant products such as red wine, anthocyanins may be actually incorporated in the structure of the tannin.

The complex problem of the structure of the condensed tannins is important because it determines their properties which, as has been noted, are linked closely to the size of the flavolan. If one could determine the degree of polymerisation by analytical means, it seems probable that one might interpret the transformations which these tannins undergo in plants and apply such knowledge in plant and food chemistry.

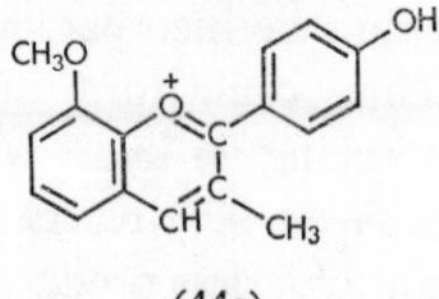

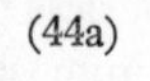

(44a)

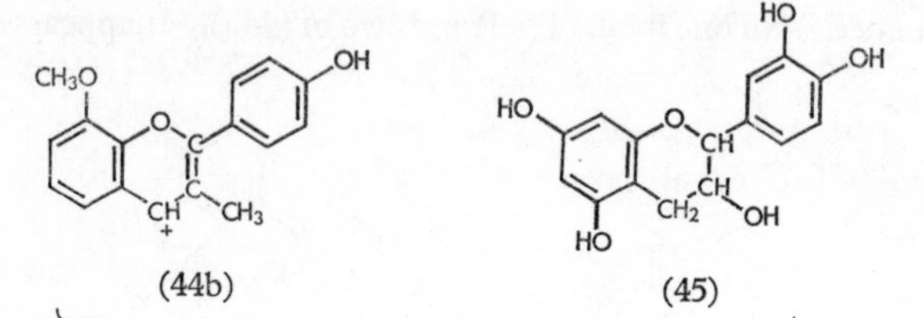

(44b) (45)

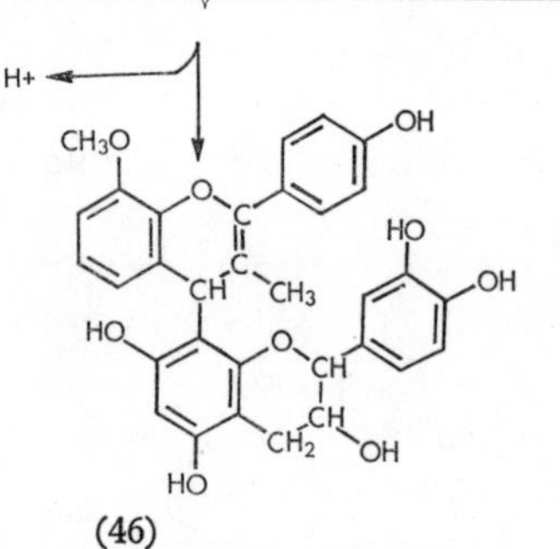

(46)

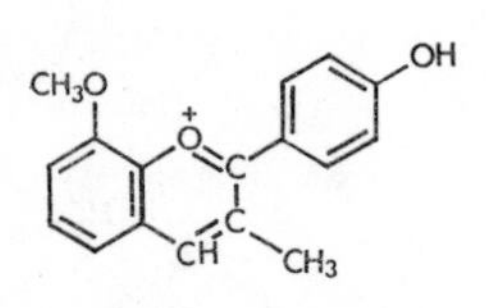

(44a)

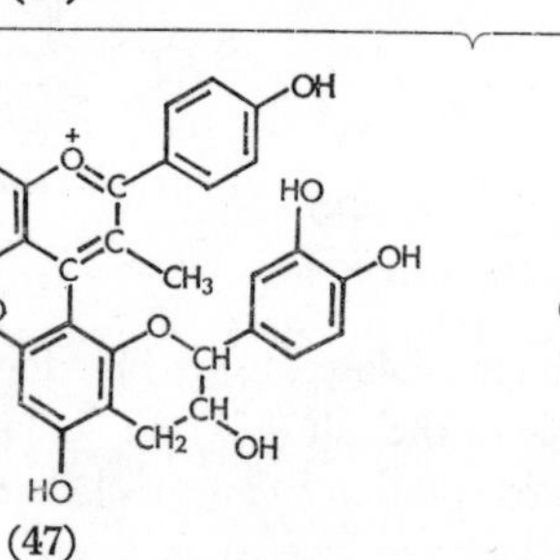

(47)

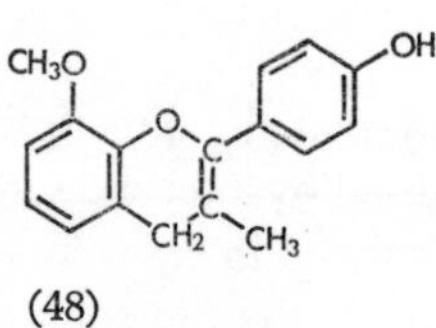

(48)

Methods for studying condensed tannins

7.9. *Introduction*

When one wishes to study tannins in a sample, the first problem which needs to be resolved is obviously how to determine the total quantity present. Methods such as oxidation with cold permanganate or phosphomolydic acid (Folin-Denis reagent) determine all phenolic compounds. However, now that the preponderant role of the flavan-3,

4-diols as precursors of the condensed tannins is known, one can use the fact that they yield anthocyanidins in hot mineral acid to determine the amount of tannins present. Nevertheless, for the analysis to be complete, it is also necessary to attempt to determine the molecular size of the tannin, since this is an important fundamental property as described in the preceding section. It is certain that every sample of a natural tannin contains several different flavolans of varying molecular weight and it is the relative proportions of each of these which will determine the overall properties of the tannin mixture. This means that some sort of fractionation will be necessary when extracting tannins from plants.

A solution to this last problem has been sought for a good many years and until recently the only practicable method was the use of hide powder which separated phenolic compounds into 'tannins' and 'non-tannins'. 'Tannins' were those substances which had a sufficient degree of polymerisation to combine strongly with hide powder. The actual method was not easy to use in a reproducible fashion and furthermore gave no indication of the size of the molecules which are separable, the essential part of the problem. The same difficulties arise if one uses, instead of hide powder, precipitation of the tannins with either gelatin or an alkaloid (cinchonine sulphate), or by saturation of the solution with sodium chloride (Masquelier *et al.*, 1959).

In the light of recent research, one might suggest that to resolve the problem of the structure of condensed tannins in an extract, the following approaches could be tried:

(*a*) fractionation of different constituents in a tannin extract by chromatographic or other means;
(*b*) selective extraction of the different tannin constituents from the tissue; or
(*c*) analytical determination of some index (e.g. V/LA or V/FD) which gives a value in direct relation to the degree of condensation.

None of these approaches has, so far, been solved in a completely satisfactory way. But having drawn attention to them, one can now appreciate better the description of the best methods presently available.

Several attempts have been made to determine the molecular weight of tannin including elevation of boiling point (Quesnel, 1968) using methylated or acetylated derivatives, ultracentrifugation (Feeny and Bostock, 1968), colorimetry (Rollman *et al.*, 1966), filtration on molecular sieves (i.e. on Sephadex, Diemair and Polaster, 1967; Lewak, 1968). The volume of solvent required to elute a constituent from a column of Sephadex is approximately a logarithmic function of its molecular weight. The higher the weight, the quicker the component will be eluted, and conversely substances which are slowly eluted have a lower molecular weight as these can penetrate the interior of the gel. By using gels of different porosity one can obtain results with varying ranges of molecular weight.

7.10. *Total amount of condensed tannins*

The determination of tannins using the 'leucoanthocyanin' reaction was probably used for the first time by Laborde in 1910 for the different parts of the grape vine and for wine. There are two possible ways of carrying out the reaction; either in aqueous (Tayeau *et al.*, 1951; Masquelier *et al.*, 1959) or alcoholic (Pigman *et al.*, 1953; Swain and Hillis, 1959) solution. The latter gives a better yield of anthocyanidin (Fig. 21), but the use of aqueous acid gives more easily reproducible results (Ribéreau-Gayon and Stonestreet, 1965b). A drawback in both cases is the fact that no satisfactory standard can be used for comparison which exactly matches the tannin under study.

Determination in aqueous solution (Masquelier *et al.*, 1959) Two tubes are prepared containing 6 ml of concentrated HCl, and one adds to each a sample of the tannin (50-150 μg) in water. One of the tubes is heated for 45 minutes under reflux at 100°, cooled under running water in the dark and 1 ml of ethanol added to dissolve any phlobaphene formed. The absorptivity is measured at 550 nm using the unheated tube as a blank. The absorptivity lies between 0·05 and 0·20 units but the reproducibility is not very good. Masquelier (personal communication) suggests that the reproducibility could be improved by extracting the anthocyanidins formed by amyl alcohol and determining their absorption in the alcoholic phase. With 50-150 μg of tannin a good straight line is obtained.

Determination in alcoholic solution Pigman and his co-workers (1953) used isopropanol-HCl as the solvent, but the relatively low boiling point of the alcohol (below 100°C) meant that special precautions were required to avoid evaporation. Swain and Hillis modified the method by using *n*-butanol and described the precautions required to ensure reproducible results. Their method is as follows: 1 ml of the solution to be analysed (which should not contain more than 50% ethanol or methanol) is added to each of two equally-sized tubes closed by a ground glass stopper. 10 ml of 5% conc. HCl in *n*-butanol is added and the contents well mixed. One of the tubes is heated at 100°C for exactly 40 minutes, the stopper being firmly closed after 4 minutes, cooled in running cold water for 4 minutes and the difference in absorptivity between the two tubes measured at 550 nm.

Govindarajan and Mathew (1965) have suggested the use of ferric iron to increase the transformation to anthocyanidins. Their reagent contains 40% concentrated HCl in *n*-butanol with 77 mg of $FeSO_4 . 7H_2O$ per 100 ml. In this case, 5 ml of reagent plus 0·5 ml of the test solution is heated at 100°C for 15 minutes.

Preparation of reference substances Condensed tannins, based on flavan-3,4-diols, can be prepared as reference substances for analysis in the following way (Masquelier, personal communication). The plant tissue is freed from lipids by treatment with light petroleum or ether and then extracted with ethanol in the cold over several days. The alcohol is removed under reduced pressure at low temperature and the residue dissolved in hot (80°C) water (equivalent in volume to the alcohol used for extraction). Insoluble matter is filtered off and the filtrate saturated with NaCl (35 g/100 ml). The precipitate of highly condensed phlobaphenes is removed and the filtrate acidified (to 1% HCl) and extracted with *n*-butanol two or three times. The tannins are displaced from the solvent by the addition of benzene (5-6 volumes) into 1% HCl (50-100 ml). This solution is again treated with NaCl and the tannins extracted with *n*-BuOH. In this case, the solvent is dried (sodium sulphate) and the tannins finally precipitated with ether. The composition of the material obviously depends on the original source, but it appears that the transformation to anthocyanidins is approximately the same irrespective of the source used.

7.11. *Fractionation of condensed tannins*

Samples of tannins are mixtures of oligomers built up from 2 to 10 units. Each oligomer has different properties and it would be useful to be able to fractionate the mixture and determine each component separately.

It might be thought that because of their relatively low molecular weight, they might be separable by paper chromatography using butanol–acetic as solvent and revealing the individual spots with vanillin (7.7). Such a method, however, only gives ill-defined streaks with the major part of the flavolans on the start line. Only the least polymerised substances migrate, but even they give poorly separated spots in most cases. The difficulty of separation is due to the affinity of the polymerised flavans and the cellulose which, as in other cases, involve hydrogen bonding between the two polymers.

Roux and Evelyn (1958a) have used chromatography in butanol-acetic on thick paper on a preparative scale to investigate the variation in R_f of the tannins with molecular weight. They cut the paper into equal-sized bands and eluted each separately. The molecular weight of each fraction was then determined ebullimetrically after methylation of the OH groups. The results showed that at R_f 0·1 the substances had a molecular weight of the order of 2,000 corresponding with a polymer containing 7 flavan monomers; R_f 0·2 corresponded with an M.W. of 1,500 and R_f 0·3 to 0·4 to a value of between 1,300 and 1,000. The monomeric forms of the flavan-3,4-diols had R_f values 0·5 to 0·7 (Roux and Evelyn, 1958b).

A second possibility for separating tannins is the use of differential solvent extraction. Ethyl acetate appears to be useful in this respect. A better resolution may be obtained by the Craig counter-current method. However, with this technique (Roux (1958b) found that although one could work with relatively larger amounts of material the separations obtained with amyl alcohol-acetic acid-water (4 : 1 : 5) were no better than by paper chromatography.

Somers (1966) successfully separated the tannins in red wine into two fractions using Sephadex G-25. Absorption was prevented by preparing the gel in aqueous alcohol and using the same solvent as eluant. The gel then functions as a molecular sieve and gives a good separation of different polymers which could be recovered quantitatively. He used a column 1·5 × 30 cm of Sephadex G-25-Fine swollen overnight in 60% ethanol containing 0·1% HCl. A paper disc was put on top of the gel to prevent its surface being disturbed. The tannins in 20 ml of red wine were precipitated with basic lead acetate and the precipitate redissolved in 1 ml of the eluting solvent. The sample was placed on the column and eluted at 12 ml per hour. The tannins separated into two well-defined bands which were recovered at the base of the column.

7.12. *Extraction of different constituents of tannins from plant material*

Hillis and Swain (1959) showed that the polymeric flavans could be divided into three categories depending on their extractability from plant tissue. For example, from plum leaves, the following three fractions were obtained:

(*a*) A first fraction which was composed of all compounds extracted by absolute methanol. This contains all the simple phenolic compounds including the lower oligomers of the flavolans all of which migrate on paper chromatography (7.11). Hillis and Swain (1959) extracted this fraction by several treatments with boiling methanol.

(*b*) A second fraction was composed of the substances not extractable by hot absolute methanol, but which were soluble in 50% aqueous methanol. This fraction was more highly polymerised and showed no mobility on paper chromatography. The components of this fraction are not necessarily insoluble in absolute methanol, but in the plant are linked to polysaccharides in the cell walls or to proteins by hydrogen bonds: bonds which can only be broken when some water is present in the extracting medium. Hillis and Swain used three extractions with boiling 50% methanol on the residue from the first fraction.

(*c*) The third fraction, which was not extractable by the above methods contained tannins more firmly bound to other cellular constituents. They probably correspond to high molecular weight polymers which are not true tannins.

By determining the total tannins on the basis of the transformation of flavan-3,4-diols to leucoanthocyanins in each fraction, it is possible to determine the proportion of tannins which are less polymerised (1st fraction) to those which are more polymerised (2nd fraction).

7.13. *Determination of the V/LA ratio*

Goldstein and Swain (1963) developed a method for determining the degree of polymerisation of tannins which consists of determining the so-called V/LA index. This determination, which only applies to tannins formed from flavan-3,4-diols, consists of applying two different analytical chemical reactions to the sample, which are affected differently by the degree of polymerisation. These are the transformation of the flavan-3,4-diol moiety to an anthocyanidin (LA) and their combination with vanillin (V) to give a coloured product. The reaction of vanillin with the phloroglucinol ring (A) of flavans has already been described (2.9) and leads to the formation of a red coloration, the intensity of which can be readily determined. Goldstein and Swain (1963) showed that the reaction is approximately stoichiometric, vanillin reacting with the 6- or 8-position of the flavan which are activated by the neighbouring OH groups. On the other hand, the linkages in the condensed tannins involve these reactive positions (38 to 42) and hence any monomeric flavan which is part of such a polymer becomes unreactive to vanillin. Thus, vanillin gives a stronger reaction (on a weight basis) with monomeric flavans than with condensed tannins. The total tannins can be determined with the leucoanthocyanin reaction although little is known about the influence of polymerisation on the yield of anthocyanidins. Roux and Paulus (1962) showed that the yield of anthocyanidin, fisetinidin, was 3·5 times greater from the monomer than the trimer. However, beyond the trimer there was little change in yield.

The conclusions of Goldstein and Swain (1963) can be summarised as follows:

(1) In the less polymerised fractions (extracted by absolute methanol, 7.12) any increase in polymerisation results in a more rapid reduction in the value of LA than of V; the ratio of V/LA is thus increased.

(2) In the more condensed extractives (those soluble in 50% methanol), any increase in polymerisation has little effect on LA but reduces the value of V; the value V/LA is thus reduced. This second observation is by far the most frequent among food products and particularly in the case of wine.

The variations in the values of LA and V as a function of polymerisation have not been demonstrated, and it should also be stressed that the method does not take into account the catechins (flavan-3-ols). Catechins react with vanillin and are thus included in any measure of V; they do not, however, give any addition to the measure of LA. Nevertheless, it must be admitted that the catechins play a less important role than the flavan-3,4-diols in the formation of most plant tannins, so that the ratio V/LA is of importance in the majority of instances.

Methods used The determination of LA is carried out as previously described (7.10). For the vanillin reaction, the method described by Hillis and Swain (1959), using a freshly prepared solution of 1% vanillin in 70% concentrated sulphuric acid, is the most useful. Four tubes are prepared, two containing the tannin to be analysed made up to 2 ml with water (no more than 0·1 ml methanol or ethanol should be present: e.g. 8% wine in water), and two containing 2 ml distilled water alone. The reagent (4 ml) is added to tubes 1 and 3 and 70% sulphuric acid to tubes 2 and 4. Using the last tube (water and acid alone) as an absolute blank, the absorptivities are measured at 500 nm in 1 cm cells.

Then the vanillin value V is given by $A_1-(A_2+A_3)$ where A_1 is the absorptivity of tube 1 (test+reagent), A_2 that of tube 2 (water and reagent) and A_3 that of tube 3 (test and acid). The ratio V/LA is a ratio of the absorptivities obtained as described.

Lewak (1968) has suggested that it is preferable to use the V/FD ratio instead of V/LA, where FD is the measure of total phenols determined by the Folin-Denis reaction (2.6, 2.16). The values of this ratio were found to correlate well with the retention time of the constituents on Sephadex LH 20 where the higher molecular weight components are eluted first (7.9). No such correlation was found using V/LA.

7.14. *Applications*

A good example of the application of the methods described above is given by the study on fruit ripening by Goldstein and Swain (1963). The tissues were extracted by methanol followed by 50% aqueous methanol and the V/LA ratio determined in each. The principal conclusions were that in the course of ripening, methanol-soluble tannins diminish and aqueous methanol soluble tannins increase while the V/LA ratios moved in the opposite direction. This showed an increase in tannin polymerisation and Goldstein and Swain (1963) suggested that the astringency of the tannins in unripe fruits is due to the presence of polymers of intermediate size which are believed to give rise to maximum astringency. On ripening, the increase in polymerisation results in a reduction of astringency.

The method has also been applied to the changes in grape tannins during ripening (Ribéreau-Gayon and Stonestreet, 1965b). It was found that the tannins in the different parts of the grape berry (pips, skin, etc.) do not have the same composition. On the other hand, treatment of wine with gelatin which removes the high molecular weight materials gives an increase in V/LA, while removal of the lower molecular weight compounds with ethyl acetate gives a lower V/LA ratio in accord with theory.

8

Metabolism and Biological Properties of Phenolic Constituents

Introduction

In considering the role which flavonoids play in plants, Lebreton and Meneret (1964) stated that 'flavonoids have received the attention of phytochemists and botanists for a long time; their abundance in vascular plants suggests that they play a part in biological processes, though this has not yet been confirmed—despite the diversity of research efforts. Do these constituents act as sexual hormones, as growth factors, as sensitisers in photosynthesis or other photobiological phenomena, as intermediates or regulators of oxidation-reduction phenomena, as constituents of cell membranes or organelles, as bactericides or fungicides . . . or are they merely secondary metabolites for storage or excretion of aromatic rings, true "by-products" whose function is no more, if one dares say it, than to favour pollination and give the poets and writers on systematics something to write about?'

This quotation shows the difficulty of considering the problem of the functions of these phenolic constituents since there are, as yet, few definite answers. For details of the results that are available, the reader is referred to works of Roubaix and Lazar (1960) and Van Sumere (1960) concerning the inhibition of seed germination; of Martin (1958), Cadman (1960), and Cruickshank and Perrin (1964) concerning the pathological functions of phenolic compounds in plants, in particular, their potential toxicity and their possible role in certain plant diseases of viral and fungal origin. The functions—both physiological and pharmaceutical—of phenolic constituents in animal organisms (Fairbairn, 1959; Ramwell *et al.*, 1964), their role in vascular resistance mechanisms (vitamin P activity) (Lavollay and Neumann, 1959), and their bactericidal action (Masquelier, 1959) are also more fully discussed in the references given.

In this chapter, discussion of the function of phenolics is limited to areas where precise data are available: their role in plant coloration and their use in genetics and taxonomy. The first section deals with the biosynthesis of phenolic compounds, particularly the flavonoids. This problem has been extensively studied in the last ten years and we can consider the essential facts well proven. In this chapter, the formation of the phenolic acids and the flavonoids alone are considered, since the formation of the tannins has already been covered in Chapter 7.

A comprehensive study of the metabolism of these substances should also include some note of their degradation. There is already some information concerning micro-organisms, and a mechanism for the degradation of the aromatic ring has been suggested by Towers (1964). In the case of the higher plants, the most interesting work is that of Zaprometov (1959) who showed that when tea leaves are fed catechins uniformly labelled by ^{14}C, they produce $^{14}CO_2$. Also, it has been shown (P. Ribéreau-Gayon, 1960) that keeping grapes for several days, at 35°C, leads to a slight diminution in the amount of anthocyanin, suggesting degradation of the substances through respiration. None of these experiments allows us to conclude categorically that plant cells are able to degrade the benzene rings in flavonoids. Moreover, we know some enzymes which degrade complex phenolic compounds, for example oxidases (2.6) and various glycosidases (5.10, 6.15), like anthocyanases, which decolorise anthocyanin solutions by hydrolysing them to unstable anthocyanidins. These reactions take place without disrupting the benzene ring.

Biosynthesis of the phenolic constituents

8.1. *Introduction*

For a long time, the problem of the biosynthesis of phenolic compounds was tackled in a purely descriptive way. From a consideration of the various substances present in a plant, and from their possible transformations in the laboratory, hypotheses were deduced for their synthesis in the cell. It was in this way that Robinson expounded his classical theory (1936) regarding the biosynthesis of flavonoids. He noted that the substitution patterns of the two benzene rings of the flavonoids are fundamentally different, and suggested that the two rings have different biogenetic origins so that the C_6—C_3—C_6 molecule of the flavonoids results from the condensation of phloroglucinol (1) with one unit of phenylpropane (2) to give a precursor (3), from which the various types of flavonoids could be derived. Though we now know that phloroglucinol itself is not involved, the essentials of this scheme remain valid, as it has been proved that the two benzene rings of the flavonoid molecule are indeed formed by entirely different routes (8.2, 8.3).

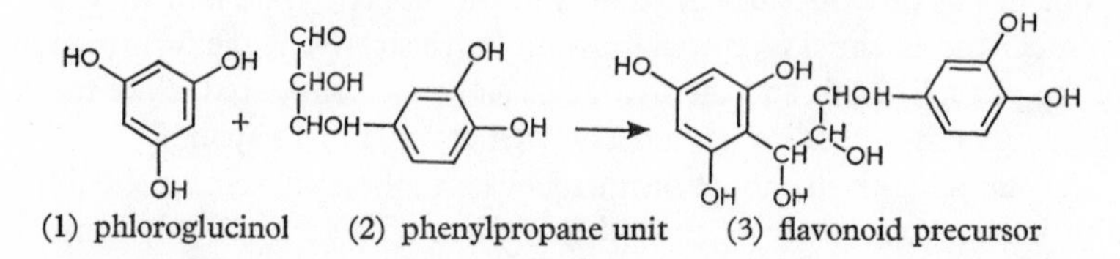

(1) phloroglucinol (2) phenylpropane unit (3) flavonoid precursor

Genetic and phylogenetic considerations have also contributed to our knowledge of biosynthetic pathways. An interesting example is Geissman and Harborne's study (1955) on the phenolics of *Antirrhinum majus* (snapdragon). There is an albino mutant of this species which, unlike normal plants, has flowers which lack flavonoids; instead, they accumulate the esters of the cinnamic acids. We can presume from this fact that the mutant has lost the capacity to condense the C_6—C_3 unit with a C_6 moiety. The C_6—C_3 units thus accumulate as cinnamic acids and are subsequently transformed into esters.

At present, the biosynthesis of cellular constituents is studied by means of three biochemical techniques:

(1) Use of mutants, which have lost their capacity for synthesising some essential substance on the pathway.

(2) *In vitro* study of the elementary biochemical reactions in the route of biosynthesis, using enzymatic systems isolated from plant materials.

(3) Use of molecules labelled with isotopic carbon (^{14}C), often in conjunction with the preceding methods.

The use of mutants, obtained artificially from a wild strain, is widespread in microbiology. New strains can be obtained which have lost their ability to synthesise some essential constituents, because one of the steps in the biosynthetic pathway has been blocked by the absence of the appropriate enzyme. Usually the metabolism is completely disrupted; however, in certain cases, the intermediate, formed just before the blocked step, accumulates. In this case, the formation of end-products which have disappeared can be re-established by adding to the culture medium the substance which is thought to be the precursor occurring just after the blocked step whose synthesis has been inhibited.

In higher plants, the use of mutants has been more limited, despite the example of *Antirrhinum majus* given above. Tissue culture (Gautheret, 1949), which often shows variations in metabolism over the intact plant, could be used and experiments on the biosynthesis of phenolic constituents have been carried out using this technique (Slabecka-Szweykowska, 1955; Paupardin, 1965, 1967; Fritig *et al.*, 1966).

The *in vitro* study of the enzymes involved in the biosynthesis of phenolic constituents, with the help of active systems isolated and purified from plant material, is in its infancy and we can hope that this method, coupled with the use of radioactive tracers, will be further developed. In the case of the phenolic constituents, this approach has been tackled by Conn (1964). However, the isolation of enzymes from plant materials is often particularly difficult owing to the complex chemical composition of plants, in particular the presence of large quantities of organic acids and tannins, both of which inhibit enzymic activity.

The use of ^{14}C-labelled molecules has given the most important results so far. As might be expected, this method has met with the greatest success in its application to micro-organisms. For, in effect, these are homogeneous cultures of cells in a physiologically well determined and reproducible state, and labelled molecules can be introduced into the culture-medium in a manner which does not disrupt normal physiological phenomena. It is thus from studies in micro-organisms that the principal results concerning the biosynthesis of the phenolic compounds have been obtained. The same pathways have been detected in higher plants, but here the use of radioactive tracers presents some difficulties. The introduction of a precursor often takes a long time and may be affected by different factors that are difficult to control. Secondary reactions also frequently appear which lead, except in short-term experiments, to a dilution of radioactivity in the end-product which makes the interpretation of the results difficult. Only labelled carbon dioxide, which is of little interest in the biosynthesis of secondary products, can be added to a plant without upsetting its normal physiological status. In all other cases, the ^{14}C-labelled substances are introduced in solution by absorption at a cut surface, usually to a detached plant organ, conditions which are very different from normal metabolism. In spite of these reservations, the use of labelled molecules has contributed greatly to the study of the biosynthesis of the phenolic compounds (see e.g. Swain, 1962b; Birch, 1962, 1963; Geissman, 1963; Neish, 1964; Grisebach, 1965; Bu'Lock, 1965; Haslam, 1966; Bilkey, 1966; Grisebach, 1967).

8.2. *Biosynthesis of the benzene ring with shikimic acid as an intermediate*

The formation of aromatic compounds from carbohydrates, with shikimic acid as an intermediate, was demonstrated by Davis (1955) by studying the synthesis of aromatic amino acids in certain micro-organisms. Wild strains of *Escherichia coli* can grow in a basic medium lacking these amino acids, which are synthesised by the cells. Davies obtained various mutant strains by ultraviolet irradiation of wild-type cultures, one of which had lost the ability to synthesise five aromatic compounds (phenylalanine, tyrosine, tryptophane, *p*-aminobenzoic acid, *p*-hydroxybenzoic acid). All these compounds had to be added to the culture medium for satisfactory growth to occur. Davis concluded that this particular strain had lost the capacity to synthesise a common precursor to all five aromatic constituents and that the addition of this precursor to the culture medium should re-establish this capacity.

After trying a great many different compounds, Davis showed that shikimic acid was the essential component. The use of radioactive tracers allowed the intermediates taking part in this mechanism to be pin-pointed (Fig. 22) and the majority of the enzymes participating in the reactions have been studied (Conn, 1964). The mechanisms of the various steps are not yet wholly understood: in particular, the cyclisation of 2-keto-3-desoxy-7-phosphoglucoheptonic acid to 5-dehydroquinic acid; and the condensation of shikimic and phosphoenolpyruvic acids which leads to prephenic acid via chorismic acid as an intermediate (3a).

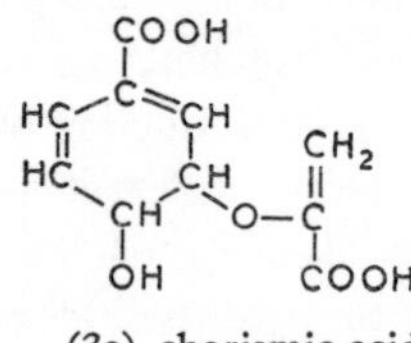

(3a) chorismic acid

The problems relating to shikimic acid and its role in metabolism have been reviewed by Bohm (1965). The reactions shown in Fig. 22 correspond to the results obtained in micro-organisms. In the case of higher plants, studies have been more limited because the appropriate mutants are not available; however, everything suggests that in vascular plants the aromatic nucleus of phenylalanine and related compounds is synthesised from glucose in the same way as in micro-organisms, i.e. *via* shikimic acid. Not only are quinic and shikimic acids widespread in the plant kingdom, but the transformation of shikimic acid into phenylalanine and tyrosine (which then give the corresponding cinnamic

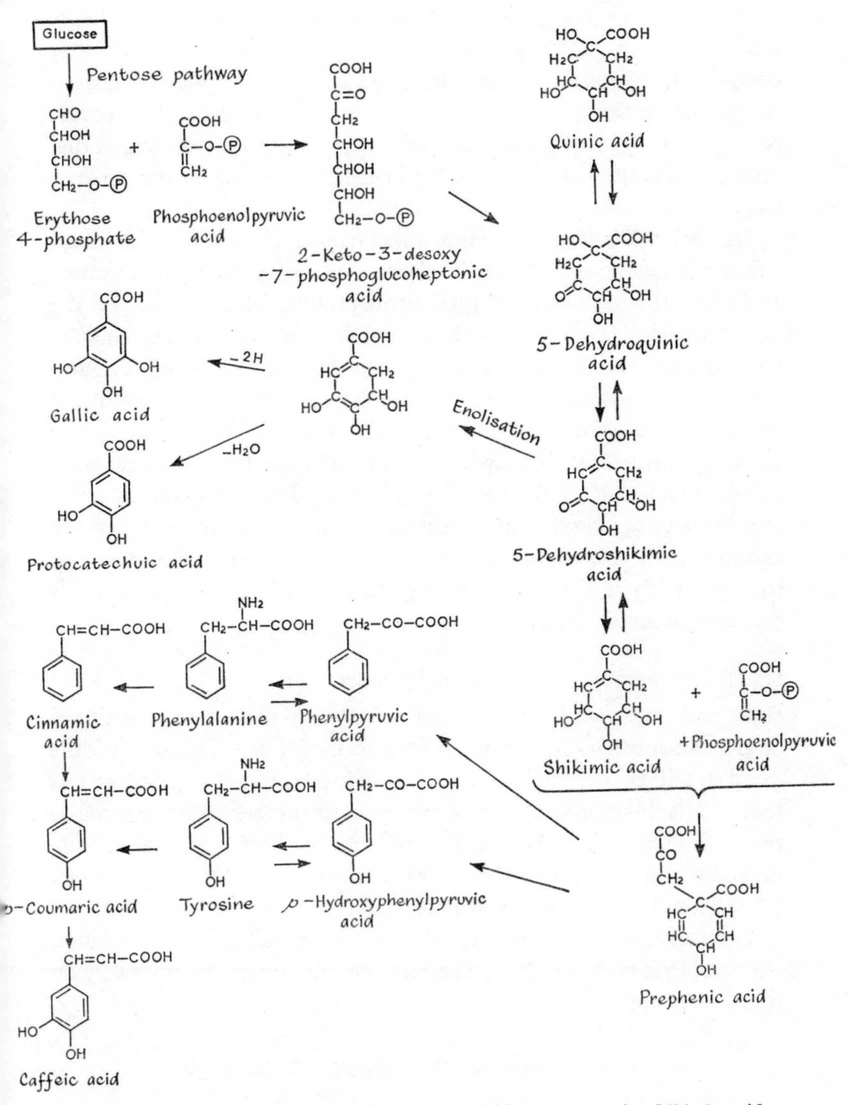

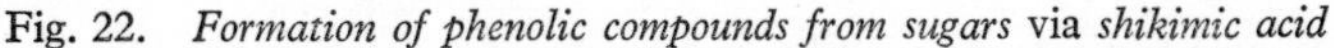

Fig. 22. *Formation of phenolic compounds from sugars* via *shikimic acid*

acids) has also been demonstrated. It should be noted that, according to the scheme (Fig. 22), 5-dehydroquinic and shikimic acids are intermediates in the pathway, but quinic acid itself is not. This latter compound, a normal constituent of plant tissues, appears to be a by-product of the principal pathway. However, Boudet and Colonna's research (1968) on quinic and shikimic acids in the oak leaf suggest that quinic acid could take part directly in the biosynthesis of the benzene ring.

The shikimic acid pathway represents the principal mode of accumulation of phenolic compounds in plants. Not only the aromatic amino acids, but the benzoic acids (gallic and protocatechuic acids) and the cinnamic acids (Fig. 22) are all formed in this way. Coumarins are derived from the cinnamic acids by a cyclisation, involving the non-enzymatic transformation of the natural *trans* cinnamic acid into the *cis* isomer (Edwards and Stoker, 1967), and the benzoic acids are probably formed by β-oxidation of the corresponding cinnamic acids (El-Basyouni *et al.*, 1964; Grisebach, 1967). The reduction of ferulic acid leads to coniferyl alcohol, which is an important precursor of the lignins (1.15). Finally, it can be seen that the side aromatic B ring of the flavonoids is formed by this pathway, with the cinnamic acids probably acting as intermediates.

8.3. *Biosynthesis of the benzene ring from acetate*

Birch and Donovan, in 1953, were the first to confirm the possibility of an aromatic nucleus being formed from acetate units. This involves the 'head to tail' condensation of acetate in which the methyl group of one molecule is linked with the carboxyl group of another with the elimination of water. The mechanism of the reaction was conclusively proved by Grisebach (1957) and has been confirmed by many authors; it is related to the synthesis of fatty acids.

The condensation of acetate units can be classified as an acylation reaction; that is to say the replacement of one methyl hydrogen by an R—CO group:

$$2\,CH_3-COOH \longrightarrow CH_3-CO-CH_2-COOH + H_2O$$

Such reactions are well known in organic chemistry, and take place most readily when the H to be substituted is attached to either an atom of oxygen (alcohol or phenol) or an atom of nitrogen (amine); the formation of an ester is an example of an acylation reaction. We also know of acylation reactions directly to carbon (C-acylation) when the hydrogen atom is sufficiently labile. An example is the self-condensation of aliphatic esters, having a hydrogen atom made labile by the presence of the α-carbon of the CO group (Claisen reaction); ethyl acetate thus yields ethyl acetoacetate $CH_3COCH_2COOC_2H_5$.

Many acylation reactions also occur in biochemistry. They usually involve thiol-ester derivatives of acids with coenzyme A (written as HS—CoA which shows the essential functional thiol group) of the type CH_3—CO ~ S—CoA, in which the high energy bond (written ~) allows the transfer of the acetyl (or other acyl) group. These esters (e.g. 'active acetate' also called acetyl coenzyme A) play a similar role to those of acid chlorides and anhydrides in organic chemistry. The presence of coenzyme A also enhances the lability of the hydrogens of the methyl group and allows acylation of the corresponding carbon (C-acylation). Though really more complex, the reaction below summarises the condensation of two acetyl-coenzyme A molecules.

$$CH_3-C{\overset{\displaystyle O}{\diagup\!\!\diagup}} \sim S-CoA + H-CH_2-C{\overset{\displaystyle O}{\diagup\!\!\diagup}} \sim S-CoA \longrightarrow CH_3-CO-CH_2-C{\overset{\displaystyle O}{\diagup\!\!\diagup}} \sim S-CoA + HS-CoA$$

We now know, as a result of Lynen's fundamental work concerning the synthesis of fatty acids, that acetyl coenzyme A first undergoes carboxylation to yield malonyl coenzyme A (malonic acid: COOH—CH_2—COOH). In this compound the lability of the H's of CH_2, situated between two CO groups, is greater than that of the H on the methyl group of acetyl coenzyme A (analogous with 'malonic synthesis' in organic chemistry). Acylation thus becomes easier and the product (4) from acetyl-CoA and malonyl-CoA is then decarboxylated to give acetoacetyl-CoA (5) which in turn can condense with a further malonyl-CoA molecule.

The reactions shown in Fig. 23 show schematically that three different types of product are possible from the condensation of any acid molecule (R—COOH), reacting in the form of acyl coenzyme A,

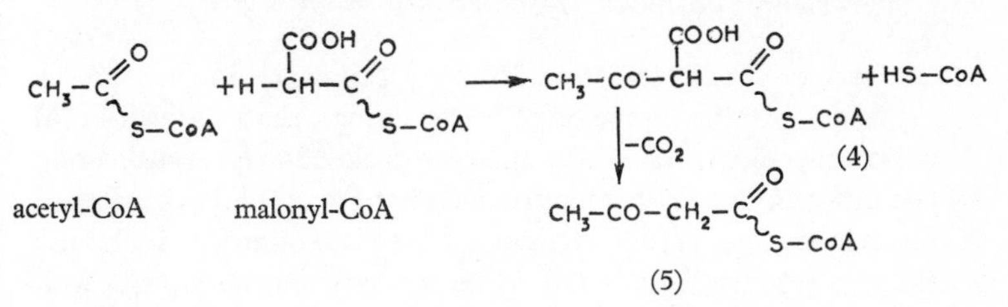

with three acetic acid molecules in the form of malonyl coenzyme A. The initial product resulting from this condensation (6) can be reduced, dehydrated and reduced again to give a fatty acid (route A); in this case, the condensation process can be continued with further molecules of

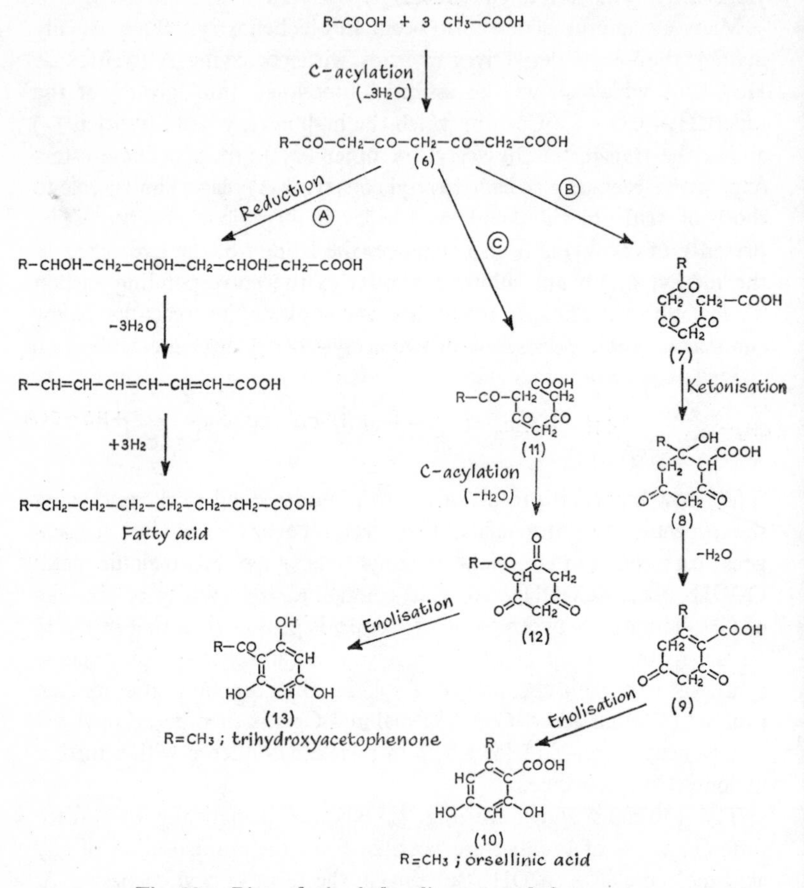

Fig. 23. *Biosynthesis of phenolic compounds from acetate*

acetate and eventually lead to a 16 or 18 carbon chain (palmitic or stearic acids). In the formation of benzene rings, the intermediate (6) does not undergo further condensation but is closed to form such a ring by two different processes (routes B and C of Fig. 23). In pathway B, (6) reacts in the form (7) and cyclisation takes place by an intramolecular aldolisation reaction, that is to say the condensation of one CO with the α-carbon of another ketone; a ketol (8) is produced, which readily loses water to give an ethylenic ketone like (9).

Enolisation of the two ketone groups in (9) gives the aromatic compound (10) (for example, orsellinic acid).

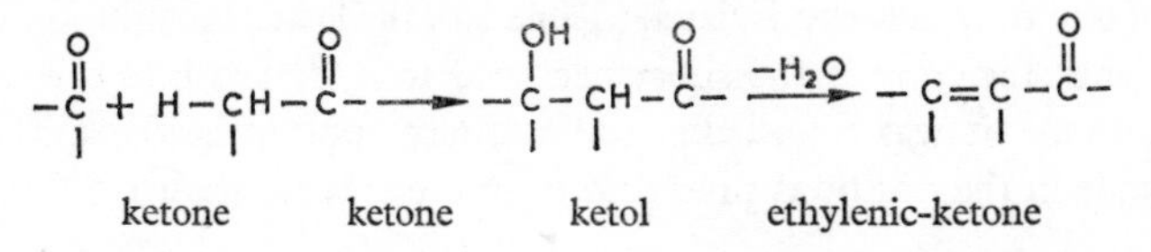

In pathway C, the condensation product (6) in the form (11) cyclises by a C-acylation reaction of the type described above between the acid group as a coenzyme A derivative and the doubly activated group. This gives the cyclic derivative (12) which, as in the preceding case, gives on enolisation an aromatic compound such as trihydroxyacetophenone (13). If the acid involved in the initial condensation is a cinnamic acid ($R{=}C_6H_5{-}CH{=}CH{-}$), cyclisation via route C leads to a chalcone having a phloroglucinol nucleus, and thence to different flavonoids (8.4).

8.4. *Biosynthesis of flavonoids*

The use of radioactive tracers has led to the acceptance of the scheme shown in Fig. 24 as the route of biosynthesis of the $C_6{-}C_3{-}C_6$ skeleton of all flavonoids. The role of the acetic acid in the formation of ring A (Fig. 24) was clearly shown by Grisebach's researches (1957) on the

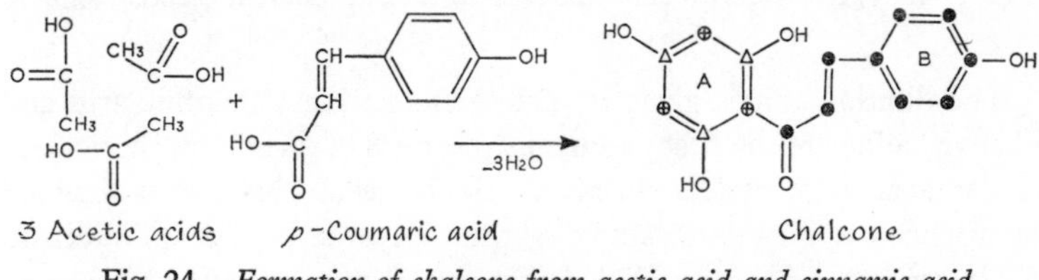

Fig. 24. *Formation of chalcone from acetic acid and cinnamic acid.*
△ = *carbons originating from acetate carboxyl*
⊕ = *carbons originating from acetate methyl*
● = *carbons originating from* p-*coumaric acid*

biosynthesis of cyanidin in red cabbage. He found that, using acetate labelled with ^{14}C on either the methyl or the carboxyl group, the distribution of the radioactivity on the carbons of ring A was in agreement with that expected from the scheme (Fig. 24).

With regard to the formation of ring B and the central three carbon moiety of the flavonoids, Underhill *et al.* (1957) found that for quercetin in buckwheat the best precursors of this part of the molecule are shikimic acid, phenylalanine, cinnamic and *p*-coumaric acids. These authors tested the efficiency of precursors by measuring the dilution of radioactivity: that is, the ratio of the specific activities of the precursor and the quercetin isolated (Table 30); the lower the dilution value the more efficiently is the substance used transformed into quercetin. In addition, it was found that radioactivity appears in the quercetin molecule in the positions predicted from the scheme shown in Fig. 24.

Table 30. *A comparison of different* ^{14}C *compounds as precursors of quercetin in buckwheat* (from Underhill *et al.*, 1957)

Compound	Dilution of radioactivity (1)
shikimic acid	36
phenylalanine	66
cinnamic acid	148, 160, 173, 338 (2)
coumaric acid	185
caffeic acid	1,268
ferulic acid	578,000
sinapic acid	4,240
protocatechuic acid	11,840
p-hydroxybenzoic acid	14,350

(1) Specific activity of the precursor/specific activity of quercetin.
(2) Values given refer to cinnamic acid labelled at different carbon atoms.

The dilution values given in Table 30 show that shikimic acid and phenylalanine are the best precursors of ring B of quercetin, the various cinnamic acids being less effective. Similar results have been obtained by Pachéco *et al.* (1965) studying the biosynthesis of delphinidin in *Viola cornuta*. They concluded that ring B was undoubtedly formed from sugars via shikimic acid as intermediate, but that the role of cinnamic acids as shown in Fig. 24 was less certain. Both Co and Markakis (1966) and Pla *et al.* (1967) showed that cinnamic acid is a better precursor of pelargonidin (IOH) and cyanidin (2OH) than is shikimic acid. On the other hand, the latter authors observed the reverse to be the case for delphinidin (3OH). They concluded that different mechanisms of flavonoid formation might exist and admitted the possibility that the synthesis of the trihydroxy B ring in anthocyanidins from shikimic acid might not involve cinnamic acid.

Chalcones probably are the only common intermediates in the synthesis of all the different types of flavonoids (Figs. 24 and 25), but this has still to be completely demonstrated. Grisebach's (1965) experimental results support the hypothesis. He showed that chalcones labelled with ^{14}C in the C_3 central moiety gave rise to labelled quercetin in red cabbage seedlings and cyanidin and quercetin in germinating buckwheat. The position of the labelled atoms in the different flavonoids was consistent with the labelling of the chalcones used. But, as Grisebach pointed out in 1965, the labelled chalcone, introduced into the plants in the above experiments may be initially degraded, and that C_6—C_3 unit released then incorporated in another form into the flavonoid molecule. However, the most recent work of the same author (1967) using double labelled (^{14}C and ^3H) chalcones seems to clearly prove the role of these compounds in flavonoid formation.

8.5. *Variation of the structure of the central C_3 moiety of the flavonoids*

Fig. 25 shows the scheme of reactions leading to the formation of the principal types of flavonoids from chalcones (Grisebach, 1965). Cyclisation of the chalcone can be easily effected by isomerisation to the flavanone which is stabilised by the formation of a hydrogen bond between the CO and OH groups in position 5 (3.4).

The flavanones and their hydroxylation products, the flavanonols (or dihydroflavonols) then act as intermediates in the formation of the different types of flavonoids. More recently, however, Wong (1968) showed that chalcones were more direct precursors of the other flavonoids than flavanones.

The reactions shown in Fig. 25 are mostly hypothetical. In most cases, however, they have been realised in the laboratory, and several other facts support the scheme.

(*a*) An enzyme capable of isomerising chalcones into flavanones has been isolated from germinating soya bean by Wong and Moustafa (1966) and its activity studied by Moustafa and Wong (1967).

(*b*) The hydroxylation of flavanones into flavanonols has been carried out by Mahesh and Seshadri (1955).

(*c*) The oxidation of flavanonols into flavonols presents no difficulty; the 2 and 3 carbons have a *trans* configuration and consequently dehydrogenation is easier than dehydration.

(*d*) Pachéco and Grouiller (1966) have chemically synthesised different flavonol glycosides from the corresponding chalcones.

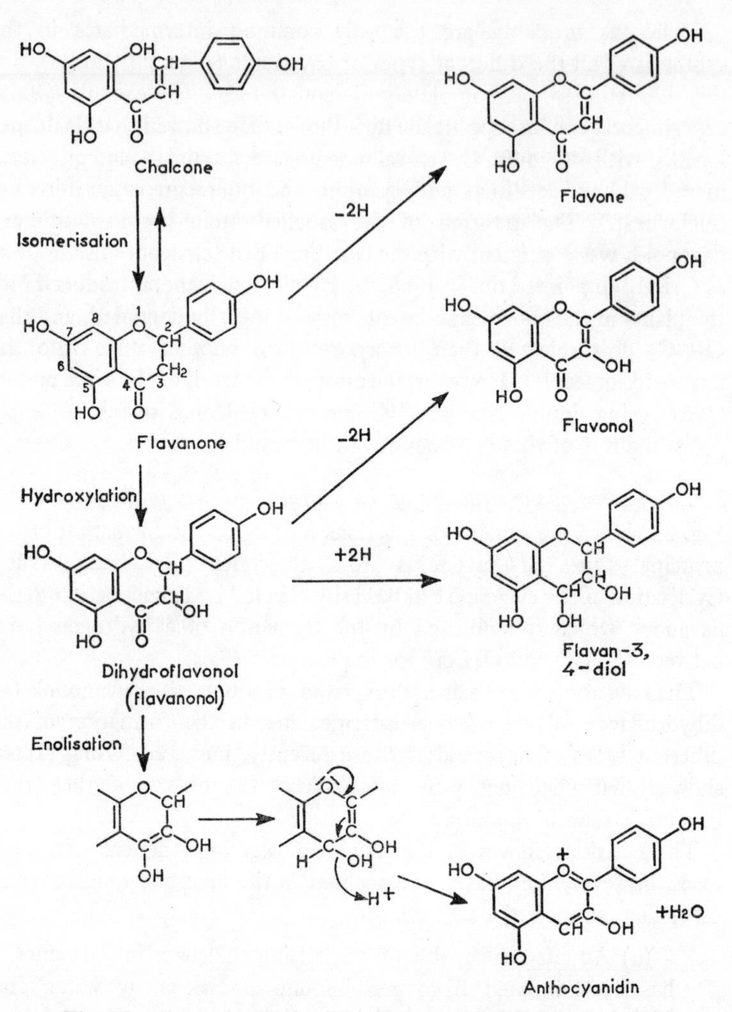

Fig. 25. *Formation of the different flavonoid classes from chalcones (after Grisebach, 1965)*

Furthermore, Wong *et al.* (1965) showed that flavanonol was transformed into flavonol in germinating chick-pea (*Cicer arietinum*).

(*e*) Pachéco (1956) has synthesised anthocyanidins from flavanonols; he envisaged the reaction passing through the flavan-3,4-diols as intermediates. Waiss and Jurd (1968), on the other hand, suggested a pathway with flav-2-enes as intermediates (1.1).

The old idea that the different flavonoids could all be derived from one another has been abandoned (e.g. the anthocyanidins by reduction of the flavonols). Thus Grisebach and Bopp (1959) showed that the feeding of labelled phenylalanine into germinating buckwheat gave labelled cyanidin and quercetin, whose specific radioactivities varied in a parallel way with time, both passing through a maximum simultaneously; this does not support the idea that one of these substances is derived from the other.

The formation of flavan-3-ols (catechins) has been studied by Zaprometov (1958, 1962), Kursanov and Zaprometov (1952) and by Comte *et al.* (1960); the results obtained are in agreement with the schemes described above and confirm the different origins of the A and B rings. The biosynthesis of isoflavones on the other hand poses a problem. One might have thought that they were formed by a different mechanism, for example from a C_6—C_2 element. However, Grisebach (1961, 1967) clearly showed that phenylalanine is a precursor of formononetin (14) in clover. This isoflavone is thus formed from the same C_6—C_3 moiety that is involved in the biosynthesis of the other flavonoids.

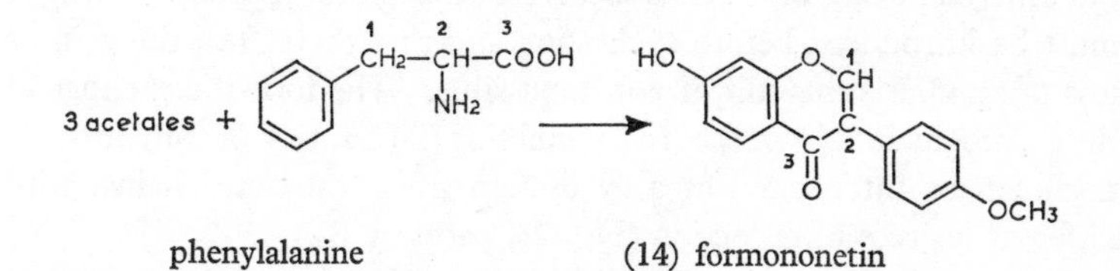

phenylalanine (14) formononetin

Moreover, the distribution of the label in the isoflavone, relative to that of the precursor, showed that biosynthesis of the isoflavonoid is accompanied by a rearrangement of phenylalanine by migration of the phenyl radical. The distribution of the carbon atoms of the precursor in the central heterocyclic ring is indicated by the numbering in (14). Rearrangements of other flavonoids to isoflavones have been obtained *in vitro* and the transformation of chalcones and flavanonols into isoflavones has been demonstrated in *Cicer arietinum* (Grisebach and Brandner, 1962; Imaseki *et al.*, 1965).

The aurones are easily obtained in the laboratory by the oxidation of chalcones, for example, sulphuretin (16) from butein (15). Furthermore, the chalcone and corresponding aurone are often found simultaneously in the same plant organ. This suggests that biosynthesis of the aurones involves chalcones as intermediates. Indeed the transformation of chalcones into aurones has been shown to occur *in vivo* by Wong (1967).

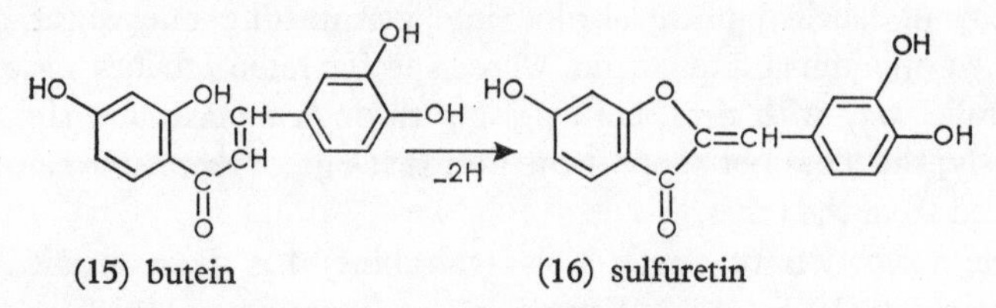

(15) butein (16) sulfuretin

8.6. *Hydroxylation and methylation of the benzene ring*

Study of the structural variation of flavonoids in the same plant family shows that the main differences lie in the number of hydroxyl groups attached to the flavonoids. This raises problems which are far from being solved.

Hydroxylation of existing flavonoids certainly depends on whether it is in the A or B ring (Fig. 24). Grisebach (1965), using labelled chalcones as precursors, showed that various isoflavones formed in plants all had identical substitution to that of the precursor in ring A, but differed in ring B. It thus seems that the hydroxyl groups of ring A must be introduced before cyclisation and that thereafter, the gain or loss of an OH is difficult, if not impossible. The formation, either of the phloroglucinol (3 OH) or resorcinol ring (2 OH, lack of 5-hydroxylation) can be interpreted as a cyclisation from triacetate chains, with different degrees of reduction (Fig. 23, pathway B).

As far as the B ring is concerned, it is generally agreed that the first OH (4′) is introduced before the formation of the C_{15} molecule of the flavonoids, perhaps even before the formation of the C_6—C_3 precursor coming from the shikimic acid pathway, although phenylalanine is a better precursor of the flavonoids than is tyrosine (Ishikura and Hayashi, 1966). But is is likely that the second OH (3′) is introduced after the formation of the C_{15} skeleton from the monohydroxylated (B-ring) chalcone. In fact we have seen (8.4, Table 30) that *p*-coumaric acid (1 OH) is a better precursor of quercetin (2 OH) than is caffeic acid (2 OH). Patschke and Grisebach (1968), in investigating the biosynthesis of kaempferol and quercetin, showed that the introduction of the second OH (3′) occurs at the stage of a dihydroflavanone intermediate (Fig. 25). Several other experiments confirmed that the second hydroxylation of the B ring occurs at the C_6—C_3—C_6 intermediate, probably a chalcone or flavanone isomer (Grisebach, 1967).

The formation of *ortho*-diphenols from monophenols in the plant is catalysed by the complex enzyme phenolase. Levy and Zucker (1960) demonstrated that a preparation of phenolase, extracted from potatoes, changes *p*-coumarylquinic acid (1OH) into chlorogenic acid (caffeylquinic acid, 2OH). Saito (1966) also showed the oxidation of *p*-coumaric acid into caffeic acid by *Saxifraga stolonifera* chloroplasts.

We have already indicated that phenolases (for example tyrosinase) have two types of activity; first they can hydroxylate monophenols into diphenols (cresolase activity) and second they oxidise diphenols into *ortho*-quinones (catecholase activity). In the plant, therefore, any diphenols produced must be protected against the second activity.

The trihydroxylated B ring (delphinidin, myricetin) can be formed by addition of an extra OH to an existing diphenol or by its direct formation from shikimic acid, analogous to that of gallic acid (Fig. 23). Pla *et al.* (1967) showed that the dihydroxylated and trihydroxylated anthocyanins in *Viola* were formed in a different way. In the trihydroxy compounds it appeared that the OH was introduced before the C_{15} skeleton was formed. *Ortho*-dihydroxyl groups might be formed similarly. Lawrence (1939) suggested that cyanidin (2OH), the most widespread anthocyanidin, was the first pigment synthesised, and was then transformed either into pelargonidin (1OH) or delphinidin (3OH).

Methylation of hydroxyl groups is most certainly carried out with methionine acting as the CH_3 group donor. Enzyme systems capable of effecting this reaction (for example the transformation of caffeic into ferulic acid) have been demonstrated in higher plants (Finkle and Nelson, 1963) and in micro-organisms (Catroux *et al.*, 1967). It is also possible that OCH_3 groups are formed by direct methoxylation of the benzene ring, without intermediate formation of the corresponding phenol (Swain, 1962b). Phenol groups could be produced by demethylation and such a mechanism would also explain the way in which *ortho*-diphenols are protected against catecholase activity.

8.7. *Formation of glycosides and esters*

Most natural phenolic constituents exist in plant as glycosides, or as esters. If a simple phenol is introduced into a plant it is quickly transformed into such a combined form, at least in living cells. When death occurs, enzymatic hydrolysis is likely to result in the appearance of aglycones. This happens, for example, in the heartwood of various trees (Hillis, 1962).

It is possible that in di- or higher glycosides, addition of sugar proceeds by stages, that is the different sugar molecules are added to the aglycone one by one (7.16). However, one cannot say in our present state of knowledge whether this addition is carried out before or after the formation of the flavonoid molecule.

In the case of the anthocyanins, one can assume that the first sugar molecule is added to position 3 before the formation of the naturally unstable anthocyanidin, which is thus stabilised (6.4). Ville and Pachéco (1967) arrived at an identical conclusion with regard to the flavonoid glycosides of *Prunus mahaleb.*

Several enzymes capable of forming glycosides from aglycones have been isolated in plants (Conn, 1964). The sugar molecule is in the form of a sugar diphosphate nucleoside, for example uridine diphosphate glucose (UDPG). Using such derivatives, the 3-monoglucoside and 3-rhamnoglucoside (rutin) were prepared from quercetin.

The most common form of combination of the phenolic acids, particularly the cinnamic acids, is as esters. Enzymatic preparations capable of catalysing the formation of such esters, have been isolated in micro-organisms (Towers, 1964) and higher plants (Corner and Swain, 1965). It is possible that this esterification takes place before the last stage of biosynthesis. From the work of Levy and Zucker (1960), Hanson and Zucker (1963), Hanson (1966), Warren Steck (1968) it appears that the principal route of chlorogenic acid synthesis in potato and tobacco leaf is: phenylalanine→cinnamic acid→*p*-coumaric acid→*p*-coumarylquinic acid→chlorogenic acid. In addition a secondary mechanism probably exists where esterification takes place at the last stage: cinnamic acid→*p*-coumaric acid→caffeic acid→chlorogenic acid.

8.8. *Physiological aspects of the biosynthesis of the phenolic constituents*

Some physiological factors, such as the influence of nutrition or light, have a considerable effect on the formation of phenolic constituents in plants (Siegelman, 1964). Thimann and Edmondson (1949) found a relation between the formation of anthocyanins and the general carbohydrate metabolism of *Spirodela oligorrhiza.* The addition of sucrose increased the synthesis of anthocyanins, while fructose was less active and glucose had no action. A similar relation was found to exist in grapes (cultivar Seyve-Villard 18 315) (P. Ribéreau-Gayon, 1959). The time, called the 'veraison' period, during which the anthocyanins are rapidly synthesised in the fruit, corresponds to a large accumulation of carbohydrates owing to a mobilisation of the reserves of the plant. Moreover, during this 'veraison' period, the ratio of monoglucosides to diglucosides is noticeably higher than during the rest of the growing period (1·08 instead of 0·47). During this time, the synthesis of the anthocyanins in the form of monoglucosides is probably more rapid than their transformation into diglucosides. The appearance of anthocyanins in some leaves in the autumn is also due to a modification of carbohydrate metabolism, namely the hydrolysis of starch (8.10).

Other observations have pointed to the influence of nutrition on the formation of flavonoids, particularly the anthocyanins. Deficiency of nitrogen favours the production of anthocyanins, variation in potassium having no action (Heller, 1948, 1948b). Copper appears to be essential and substances like phenylthiocarbamide, which form copper complexes, inhibit synthesis of anthocyanins (Edmondson and Thimann, 1950). Some antibiotics (terramycin, streptomycin, aureomycin), on the other hand, enhance the formation of anthocyanins (Nétien and Lacharme, 1955; Siegelman and Hendricks, 1957), whereas substances such as methionine and ethionine (Thimann and Radner, 1955a, 1955b) or *p*-fluorocinnamic acid (Grisebach, 1965) inhibit it. With the help of specific inhibitors, Stafford (1966) showed a relation between the synthesis of anthocyanins and of proteins.

The influence of light on the formation of the anthocyanins has been known for a long time, and in normal plants alternative light and dark periods are necessary (Siegelman and Hendricks, 1957, 1958). However, in the presence of sucrose, anthocyanin synthesis can take place in the dark (Havelange and Schumacker, 1966). In the case of grapes (P. Ribéreau-Gayon, 1959) the amount of anthocyanins formed during the maturation period is lowered by the reduction of daylight (the adsorptivity of an extract obtained under standard conditions falls from 2·7 to 0·5) as well as by night (from 3·6 to 2·1).

Siegelman and Hendricks (1957) and other authors showed that the formation of anthocyanins is controlled by two photochemical reactions; one shows a maximum activity at wavelengths of 660 and 730 nm (photoresponse II); the other maximum activity at 690 and 450 nm (photoresponse I). The photoreceptor II is well defined; it is the phytochrome which plays a role in numerous aspects of the plant's development (Butler *et al.*, 1965; Hendricks and Borthwick, 1965). Phytochrome is a chromoprotein existing in two forms—one absorbing light at 660 nm (P660) and the other at 730 nm (P730); the irradiation of P660 by a wavelength of 660 nm transforms it into P730, and vice versa. The P730 form is involved as a trigger in the synthesis of the anthocyanins.

The nature of the photoreceptor I is not known. Moreover, its mode of action seems to vary with the tissues under consideration (Siegelman, 1964). According to Vince and Grill (1966), phytochrome might be involved here also, but in another form. Creasy (1968) observed an increase in phenylalanine ammonia lyase activity by exposing discs of strawberry leaves to the light. This enzyme, whose activities parallel the appearance of flavonoids, is necessary for their synthesis.

The mechanism by which light controls the formation of phenolic constituents is unknown. It seems that the energy is utilised particularly in the condensation of acetyl coenzyme A units used in the formation of the A benzene ring. The role of the light in the synthesis of flavonols (Bottomley *et al.*, 1966; Russell and Galston, 1967) and flavans (catechins and leucoanthocyanins) (Creasy and Swain, 1966) has been studied. With regard to the leucoanthocyanins, Combier and Lebreton (1968) observed the existence of a diurnal rhythm in *Sedum album;* the concentration of the compounds is at a maximum at the end of the night and at a minimum at midday. The deviation may be up to 50% of the mean value.

The role of flavonoids in plant coloration

8.9. *Introduction*

The water-soluble flavonoids, together with the fat-soluble chlorophylls and carotenoids, are the principal pigments of plants. The flavonoids are responsible for most red, blue and violet colours of plant organs; they also contribute to yellow colours, but in this case the carotenoids are more important.

The coloration of plants is not only dependent on the chemical nature of the pigments but also on a number of physical and chemical factors which modify the *in vitro* colours of the pigments. Commoner (1948) showed that the absorption spectrum of the pigment in the cell is different from that of the pigment in solution. The intervention of these different factors explains how the multitude of hues observed in nature can be derived from a relatively few pigment structures. These aspects of plant colours have been reviewed by Sannié and Sauvain (1952), Blank (1958) and Harborne (1965b).

8.10. *Role of the anthocyanins*

The influence of structural factors on the colour of anthocyanins has already been mentioned briefly (cf. Tables 22 and 24, 6.6). The main facts are as follows:

(1) Hydroxylation of the B ring is the predominant factor; generally speaking, flowers having a red-orange or pink coloration contain pelargonidin (1OH), those which are red have cyanidin (2OH) and those which are purple or blue contain delphinidin (3OH). Further, the absence of the OH in position 3 causes a shift towards yellow colours.

(2) Methylation of the hydroxyl groups leads to a change in colour towards the red region. The shift is a small one in the case of hydroxyl groups of the B ring, so that malvidin is only slightly redder than delphinidin. This modification is more important in the case of a hydroxyl in position 7 (giving rosinidin and hirsutidin).

(3) The difference between aglycone and glucoside is a hypsochromic shift of about 12 nm; however, aglycones do not normally exist in the free state in petal tissues. Of the various glycosides, the 3,5-diglycosides have a slightly bluer colour than the 3-glycosides.

Saitô (1967) measured the absorption spectra of anthocyanins directly in the petals of about forty plants and found that the plants fell into four groups; in only one of these groups was the *in vivo* spectrum identical to that of the pigment *in vitro*. Concentration factors also influence the coloration of anthocyanins in plants; for instance, the amount of pigment in the classical varieties of cornflower is about 0·05 to 0·07% (dry weight) while some purple varieties contain 13 to 14%.

pH is known to affect the colour of anthocyanins, which are red in an acid medium and blue in a neutral or alkaline medium (6.4). The contribution of this factor, in interpreting the blue coloration of flowers, has been debated. Karrer (1928) observed that blue flowers are richer in mineral salts (9·1 to 13%) than red flowers (4 to 6%), suggesting that they have a higher pH; however, later measurements (Hayashi and Isaki, 1946) showed that the pH of the cell sap of all plants is acid. Nowadays, it is accepted that pH does not contribute towards the blue coloration of flowers and that the formation of anthocyanin complexes with metals (iron, aluminium, magnesium) (2.4, 6.5) plays the major role in this phenomenon.

The cyanidin 3,5-diglucoside complex, 'protocyanin', responsible for the blue colour of the cornflower, has been studied in particular detail (Bayer, 1958; Bayer *et al.*, 1960; Hayashi *et al.*, 1961; Saitô *et al.*, 1961; Bayer, 1966). It has a molecular weight of about 20,000 and contains a pectic polysaccharide, as well as anthocyanin, iron, magnesium, and sometimes aluminium. As early as 1933, Robinson and Robinson suggested that absorption of anthocyanins on negatively charged colloidal particles of polysaccharides, would stabilise the blue form of the anthocyanin at an acid pH. Bayer (1966) proposed a formula for such a complex. The influence of the colloidal state on the colour of anthocyanins has also been mentioned by Werckmeister (1965). Also working with the cornflower, Asen and Jurd (1967) have isolated another blue pigment, cyanocentaurin, different from protocyanin; this pigment, obtained in a crystalline form, seems to be a complex containing iron, four molecules of cyanidin 3,5-diglucoside, and three molecules of a biflavone glucoside.

The formation of metallic complexes does not explain the blue colours given by those anthocyanins which lack a free *ortho*-diphenolic group (e.g. malvidin derivatives) (6.5). Such colours are due to 'copigmentation' or to the formation of a complex between the anthocyanin and the copigment which is usually a hydrolysable tannin or flavone glycoside; this phenomenon must not be confused with the superposition of several pigments of different colours upon each other. Copigmentation occurs, for example in the rose, a plant in which a blue variety has long been sought (Harborne, 1965b); the few bluish varieties available are not wholly satisfactory and are, in fact, really mauve or purple, the flowers containing cyanidin diglucoside as in the red varieties but copigmented by a quantity of gallotannin. Jurd and Asen (1966) showed that quercetin 3-rhamnoside (quercitrin) and chlorogenic acid act at pH 5·45 as copigments with respect to cyanidin 3-monoglucoside only in the presence of aluminium.

Anthocyanins are also responsible for reddish coloration in leaves of higher plants. They can act in three ways, giving:

(*a*) permanent red coloration, limited to some ornamental species (*Prunus*);

(*b*) temporary red coloration appearing either at the seedling stage, or after tissue-wounding or an infection of viral or fungal origin; and

(*c*) autumnal red coloration, a well-known phenomenon.

Red autumnal colorations in leaves are probably related to modifications in carbohydrate metabolism, e.g. the liberation of sugar by starch hydrolysis in autumn. In all cases, leaf anthocyanins are chemically simpler than those of flowers and fruits. The commonest pigment is cyanidin in its simplest glycosidic form as the 3-monoglucoside.

8.11. *Role of flavones and related compounds*

Leaves contain different flavone pigments, but the natural colour of these pigments is usually masked by the green of the chlorophylls. In flowers and fruits, these pigments may contribute towards yellow coloration, but most yellow colours in higher plants are due to carotenoids.

Harborne (1965b, 1965c) noted that three groups of flavonoid participate in yellow coloration of plants:

(1) The chalcones and aurones, which occur, for example, in the dahlia; petals containing these pigments, when exposed to ammonia vapour, turn red.

(2) 8-Hydroxylated flavonols such as herbacetin and gossypetin (5.1), present in *Primula* and *Rhododendron* species.

(3) 3-Desoxyanthocyanidins (apigeninidin and luteolinidin) (6.1).

In addition, cream, ivory, or white flowers nearly all contain colourless flavonoids (flavones and flavonols).

Taxonomic and phylogenetic importance of phenolic constituents

8.12. *Introduction*

Botanists have always tried to produce systems of plant classification which are both phenetic and phylogenetic in content. Such classifications are continually being modified as new evidence from morphology, anatomy and palaeontology is being brought into consideration. Although chemical characters have been used for a long time for purposes of classification, chemotaxonomy as such (Heywood, 1966, 1968) has only developed significantly since the discovery of paper chromatography. For a chemical constituent to play a role as a taxonomic marker, it must fill certain requirements.

(1) It must not be a primary plant metabolite (carbohydrates, organic acids).

(2) Conversely, it must not have too specialised a structure elaborated by a limited number of species. Also it is as well if it belongs to a class of compound in which many related structures exist.

(3) It must accumulate and consequently it must be stable and be only slowly metabolised.

(4) It must be easy to detect.

Phenolic constituents that fulfil these requirements satisfactorily can play an important role in plant chemotaxonomy (Bate-Smith and Metcalfe, 1957; Bate-Smith, 1958, 1962b; Swain, 1963; Alston and Turner, 1963; Lebreton, 1964; Lebreton and Meneret, 1964; Swain, 1966; Harborne, 1967a; Bate-Smith, 1965, 1968).

With reference to the flavonoids, much information concerning their distribution in plants and their taxonomic significance has been brought together by Harborne (1967a). Further, he has discussed the contribution that the flavonoids might make in elucidating phylogenetic trends, that is the study of the origin and evolution of plant groups. Although there are some exceptions (8.13), there has been a general tendency in the course of evolution towards the elaboration of more and more complex flavonoids:

(*a*) In algae, bacteria and fungi, flavonoids are almost completely lacking; the few reports described in the literature are of atypical occurrences.

(*b*) In mosses, flavonoids are limited to a few types: flavonols, flavone *C*-glycosides (5.3), 3-desoxyanthocyanidins (apigeninidin and luteolinidin, 6.1).

(*c*) In ferns, desoxyanthocyanidins, flavones, flavonols, leucoanthocyanidins, chalcones and flavanones are present; most of the substances have simple structures.

(*d*) In gymnosperms, nearly all classes of flavonoids are present but in this group again the substances are relatively simple. Biflavonyls (1.11) are frequent.

(*e*) In angiosperms, the whole range of flavonoids is present. Furthermore, there exists a correlation between the classification (Davis and Heywood, 1963) and the nature of the pigments. The most complex structures are found in the most highly evolved families such as Leguminosae, Gesneriaceae and Compositae whereas in the more primitive families, such as Magnoliaceae and Ranunculaceae, flavonoid structures are at their simplest.

8.13. *Taxonomic and phylogenetic significance of the phenolics of the angiosperms*

Bate-Smith (1962, 1968) carried out a most extensive survey of phenolic constituents in more than a thousand angiosperm species (both monocotyledons and dicotyledons). He found that there is relation between the woody character of plants and the presence of leucoanthocyanins, that is tannins; these substances are much more frequent in woody than in herbaceous plants.

In the case of the cinnamic acids, non-methoxylated caffeic acid is widespread, more particularly in woody species, while methoxylated cinnamic acids are more frequent in herbaceous species. Since lignin is known to be formed from a methoxylated C_6—C_3 molecule (coniferyl alcohol, 1.15), the accumulation of methoxylated cinnamic acids is presumably related to inhibition of lignification.

Some other phylogenetic conclusions can be drawn from a consideration of the distribution patterns of phenolic constituents:

(1) The ability to synthesise leucoanthocyanins (flavan-3,4-diols) and flavonols, which are oxidised molecules, is a primitive characteristic; in the course of evolution, less oxidised products are formed, flavan-3-ols (catechins) in the one case and flavones in the other.

(2) The presence of a trihydroxylated benzene ring is also a primitive characteristic. It is possible that such a group is synthesised directly from a non-aromatic precursor, by a mechanism different from that leading to mono- and diphenols (8.6).

(3) The loss, in the course of evolution, of the ability to synthesise either leucoanthocyanins or trihydroxylated derivatives is an irreversible phenomenon.

Consequently, the general tendency towards the elaboration of more and more complex structures in the course of evolution imposes certain restrictions; as compared with primitive characters, advanced characters result from mutations corresponding either to the acquisition or loss of some biosynthetic function.

8.14. *Examples of the taxonomic use of phenolic constituents*

From the recent literature, examples have been selected which show a correlation between the taxonomic position of species and the nature of their pigments.

(1) All species of the apple genus, *Malus*, contain a dihydrochalcone (phloretin, 1.4), present as the glucoside phloridzin, whereas all species of the pear genus, *Pyrus*, have arbutin (hydroquinone glucoside, 1.7) instead (Williams, 1960, 1966). These rosaceous genera were once put together and their more recent separation by taxonomists is supported by the above chemical data.

For a more detailed account of the distribution of phenolic constituents in the genus *Pyrus*, see Challice and Williams (1968a, 1968b).

(2) Isoflavones (1.3) characterise the subfamily Lotoideae (Papilionatae) of the Leguminosae (Ollis, 1962). However, they do occur occasionally in plants of some other families, e.g. the Rosaceae, Moraceae and Iridaceae.

(3) Nine of the ten families constituting the order Centrospermae are characterised by the absence of anthocyanins and their replacement by other red-purple pigments, the betacyanins (Mabry, 1966; 1.12). As a result of this discovery, the affinities of the tenth family, Caryophyllaceae, which have ordinary anthocyanins, have been the subject of much debate.

(4) The biflavonyls (e.g. ginkgetin, 1.11) are rarely met with in the angiosperms; on the other hand, they are frequent in the gymnosperms excepting the family Pinaceae (Baker and Ollis, 1961).

(5) Within the family Pinaceae, flavanonols (2,3-dihydroflavonols, for example taxifolin, 1.3) are present in the genus *Pinus* but absent from the genus *Abies* (Pachéco, 1957). Furthermore, the distribution pattern of all flavonoids present (flavones, flavanones and flavanonols) confirms the division of the genus *Pinus* into two subgenera: the first, Haploxylon, trees with the needles in groups of five, have at least four flavonoids in the heartwood; and the second, Diploxylon, trees with needles in groups of two or three, have at most two flavonoids in the heartwood (Erdtman, 1957).

(6) The presence of ellagic acid is restricted to the dicotyledons; it is absent from the monocotyledons, gymnosperms and ferns. In the family Rosaceae, this acid has been found exclusively in the subfamily Rosoideae, to be precise in six of the seven tribes of this subfamily; it is absent from the seventh tribe, the Kerrieae, whose relation to the otherwise very homogenous Rosoideae is debatable (Bate-Smith, 1962b).

(7) Bate-Smith *et al.* (1968) showed that umbelliferone (4.4) is a good chemical marker for the genus *Hieracium*. Significantly, it is absent from those former *Hieracium* taxa now sometimes separated on morphological and genetical grounds as the genus *Pilosella* Hill.

(8) Anthocyanidins lacking hydroxylation at position 3 (3-desoxyanthocyanidins, apigeninidin and luteolinidin; 6.1) occur almost exclusively in the family Gesneriaceae. Their distribution

is correlated with the subfamily division; the Gesnerioideae contain these pigments whereas the Cyrtandroideae lack them (Harborne, 1967a).

(9) 5-methylquercetin (azaleatin; 5.1) is a good example of a substance of taxonomic interest at the species or genus level. It is a pigment of fairly restricted distribution, easily identified on chromatograms because of its bright yellow fluorescence. This substance, along with 5-methylkaempferol, is found in six families which otherwise have few characters in common but which are all predominantly woody (Harborne 1967a, 1969b). In the Plumbaginaceae, the presence of azaleatin follows the division of the family into two tribes; the Plumbagineae contain this flavonol and the Staticeae lack it. 5-hydroxy-2-methylnaphthoquinone (plumbagin), present in the species of the first tribe and absent from those of the second, also acts as a good taxonomic marker in this family (Harborne, 1967b). In the genus *Eucryphia* (Eucryphiaceae) which contains but five species there is a relationship between the presence of azaleatin and the geographical origin; it is in the two species of South American origin, but not in the three Australian species (Bate-Smith *et al.*, 1967).

(10) 8-hydroxyquercetin (gossypetin, 5.1) acts as a taxonomic marker of particular interest (Harborne, 1969a). In the Primulaceae, this flavonol replaces carotenoid as the yellow flower pigment in two sections of the genus *Primula*. Gossypetin is also present in two genera closely related to *Primula* (e.g. *Dionysia*) but is absent from more distant taxa (e.g. *Lysimachia*). In the Ericaceae, the same pigment is found only in some sections of the genus *Rhododendron* and in some other more or less closely related genera (e.g. *Ledum*).

(11) A very extensive survey of phenolic constituents of the leaves of *Eucalyptus* species has been carried out by Hillis (1966, 1967a, 1967b, 1967c, 1967d). The occurrences of myricetin, ellagic acid, leucoanthocyanins and other unidentified phenolics are correlated with sectional and series classification of the various species.

(12) Similarly, correlations between systematics and the nature of flavonoids in the family Primulaceae have been studied by Harborne (1968). This author surveyed about a hundred species belonging to eighteen of the twenty-five genera of this family. The nature of the anthocyanins is extremely variable depending on the

genus; the anthocyanidins present are cyanidin, peonidin, delphinidin and malvidin as well as the 7-methylanthocyanidins (hirsutidin and rosinidin, 6.1). Among the other flavonoids, leucocyanidin, leucodelphinidin and kaempferol are widespread; on the other hand, gossypetin (see section 10 above) and 3′,4′-dihydroxyflavone are limited to certain groups of closely related genera. Finally, the two flavones apigenin and luteolin have been found in only one genus *Soldanella* whose evolution, from a phylogenetic point of view, is supposed to be well advanced, on the assumption that the replacement of flavonols by flavones is an advanced character (8.13).

(13) Albach and Redman (1969) studied the distribution of flavanones in the fruits of forty-one varieties belonging to the eighteen species of the taxonomically difficult genus *Citrus*. Chemically, the substances vary according to the aglycones which are principally naringenin, eriodictyol, hesperetin (5.1) as well as their methyl derivatives, or to the nature of the sugar molecule present (rutinose or neohesperidose). Qualitative and quantitative differences in phenolic constituents are useful for chemically characterising the different species.

(14) A chemotaxonomic investigation of the order Urticales, undertaken by Lebreton (1964), led to some suggested modifications of the present classification. The pattern of distribution of flavonoids in the species of the various families confirms the heterogeneity, well known from the botanical point of view, of that order. The Urticales comprises three families (Ulmaceae, Urticaceae, Moraceae); the Moraceae is further subdivided in the traditional classification into subfamilies and tribes: Moroideae (Moreae, Broussonetieae, Dorstenieae), Artocarpoideae (Euartocarpeae, Ficeae) and Cannaboideae. The Cannaboideae (in particular the hop, *Humulus lupulus*) possess a flavone content different from that of the other Moraceae, especially by having leucoanthocyanins. This agrees with the raising of this subfamily to the rank of family, independent of the Moraceae, which has already been proposed on morphological characters (absence of latex and herbaceous habit). In the Artocarpoideae, the Euartocarpeae contain flavonoids similar to those of the Moroideae, which suggests that these tribes could be united. On the other hand, the tribe Ficeae is distinctive in containing leucoanthocyanins and might therefore be justifiably raised to subfamily level.

8.15. *Taxonomic significance of phenolics in the genus* Vitis

The distribution of phenolic constituents in the grape has been studied (P. Ribéreau-Gayon, 1959, 1964a). In the case of the anthocyanins, the results (Table 31) show primarily that within a given species (*V. vinifera*), different varieties have the same chemical composition; Lebreton and Meneret (1964) arrived at the same conclusion from studying hops. The influence of different environmental conditions was also examined. The total amount of anthocyanin can vary considerably, in particular as a function of climate (this is not shown in Table 31); by

Table 31. *Anthocyanin contents of grapes of two varieties of* V. vinifera *grown in different localities (the values are given as percentages of the total colouring matter and represent all the glycosides present of each anthocyanidin)*

	Merlot, 1957 Bordeaux vignobles				Merlot 1958	Cabernet-Sauvignon, 1958		
	1	2	3	4	Bordeaux	Bor-deaux	Mont-pellier	Cali-fornia
delphinidin	17	21	19	23	18	12	12	5
petunidin	11	15	14	14	12	12	9	4
peonidin	18	14	15	15	13	19	12	0
malvidin	54	50	52	28	57	57	67	92

contrast, the nature of the pigments and their relative proportions do not change. Within any one geographical region, the pigment composition of a variety is uniform but there is variation when the same variety is grown in different locations.

One cultivar of *V. vinifera*, Hamburg muscat, is exceptional in accumulating cyanidin derivatives rather than delphinidin-based pigments. It is interesting that Hamburg muscat is, in Negrul's (1946 classification), the only one among the cultivars studied which does not belong to the same ecologico-geographical group. Turning to the different species of the genus *Vitis*, there are significant differences in the nature of their anthocyanins; some species produce 3,5-diglucosides (*V. rotundifolia*, *V. riparia*, *V. rupestris*) while others do not. Finally, while in most species delphinidin derivatives are quantitatively important, this is not so in the case of *V. lincecumii* and *V. aestivalis*, two species which are considered, if not identical, at least as being very close to each other.

Genetics of phenolic constituents

8.16. *Introduction*

In the course of the last few years, the study of genetics has made real developments owing to the progress in understanding the gene and more particularly the structure of nucleic acids; Watson and Crick's DNA model provides a satisfactory interpretation of the mechanism of duplication of genes and the transfer of hereditary characters. By contrast, the progress accomplished by classical Mendelian genetics in higher plants is less spectacular, because they lend themselves less easily to genetical studies than do micro-organisms. However, geneticists have always been interested in the pigmentation of plant organs. On the one hand, it is a character that is easy to score and, on the other, the production of new colour forms in flowers is of obvious horticultural interest. Earlier reviews of the genetics of constituents and especially anthocyanins are those of Wheldale-Onslow (1925), Scott-Moncrieff (1936), Lawrence and Price (1940) and Sannié and Sauvain (1952). The development of chromatographic techniques gave a new impulse to this research. The many studies, carried out in the 1950s, have been reviewed by Harborne (1963) and Alston (1964). The genetics of the flavonoids of *Trifolium subterraneum* has been studied recently by Wong and Francis (1968).

The aim of such studies is to interpret, according to the laws of genetics, differences in the phenolic constituents observed in plants produced by crossing different colour varieties within a species. In the case of the flavonoids, the various natural compounds are related to each other by relatively simple structural substitutions; it has been demonstrated (Harborne, 1962a) that, in most instances, each of the (presumably enzyme-mediated) chemical reactions such as hydroxylation and methylation is associated with a particular gene.

8.17. *Genetic control of the structural modification of flavonoids*

The biosynthetic mechanisms by which each of the various types of flavonoids are produced are clearly related to each other but are difficult to interpret in genetical terms, in our present state of knowledge. Thus, the pathway leading to flavones or flavonols is certainly similar to that yielding anthocyanins. However, the synthesis of the anthocyanins is probably more complex as it requires more steps; in fact, in numerous plants, there are varieties containing only flavones or flavonols, and others containing anthocyanins in addition to the same flavones or flavonols.

Thus, the production of anthocyanins is controlled by several genes, which assumes the intervention of several biochemical reactions. For example, the crossing of two *Vitis* varieties having white grapes (Sylvaner and Riesling) results in a hybrid grape which contains anthocyanins (Professor Becker Geisenheim, personal communication). There must therefore be two genes, which are alternatively dominant and recessive in the two white varieties; crossing produces a variety in which both genes are dominant, and hence anthocyanin is produced. The view that anthocyanins are formed from leucoanthocyanin precursors which was much in vogue at one time is now discounted. For example, Feenstra (1960) showed that in *Phaseolus vulgaris*, there is a gene that uniquely controls the synthesis of leucoanthocyanins, which is independent of anthocyanin synthesis.

The most simple structural modifications in the flavonoid molecule, such as hydroxylation, methylation and glycosylation, are those that have been most studied from the genetical point of view. Hydroxylation and methylation are individually controlled by single genes, the more highly hydroxylated or methylated forms being the dominant types. The hydroxylation and methylation processes rarely go to completion, and intensity genes are also involved; in the case of the grape *V. vinifera*, in nearly all varieties, the ratio between the quantities of trihydroxylated and dihydroxylated anthocyanin is of the order of 4 : 1; in muscats, it is about 0·5 : 1. This is constant pigment character, independent of the growing conditions.

The glycosylation of anthocyanins was studied by Harborne (1962a) in *Streptocarpus*: the five genes controlling glycosylation act as follows:

$$\text{Precursor} \xrightarrow{\text{Q}} \text{3-glucoside} \xrightarrow{\text{X, Z}} \text{3,5-diglucoside}$$

$$\text{3-glucoside} \xrightarrow{\text{P}} \text{3-xyloglucoside} \qquad \text{3,5-diglucoside} \xrightarrow{\text{D}} \text{3-rhamnoglucoside -5-glucoside}$$

Q, P and D each control a well-defined step in synthesis, whereas the formation of 3,5-diglucosides from 3-monoglucosides requires two complimentary genes, X and Z. In the genus *Vitis*, the production of diglucosides from monoglucosides is controlled by only one gene (8.19) (Boubals *et al.*, 1962; P. Ribéreau-Gayon, 1964a). The above scheme shows that a hybrid might possess a complex glycoside not present in either parent (Alston *et al.*, 1965). Though there are several exceptions, the formation of di- or triglucosides is dominant to the formation of mono- and diglucosides respectively.

8.18. *Genetic control of the formation of flavonoids in* Antirrhinum majus
The inheritance of colour in snapdragon flowers has been much studied (Geissman *et al.*, 1954; Dayton, 1956; Sherratt, 1958). The results have been summarised by Swain and Bate-Smith (1962) who interpreted the formation of flavonoids by the scheme in Fig. 26. Four principal genes are involved: N, Y, M, P.

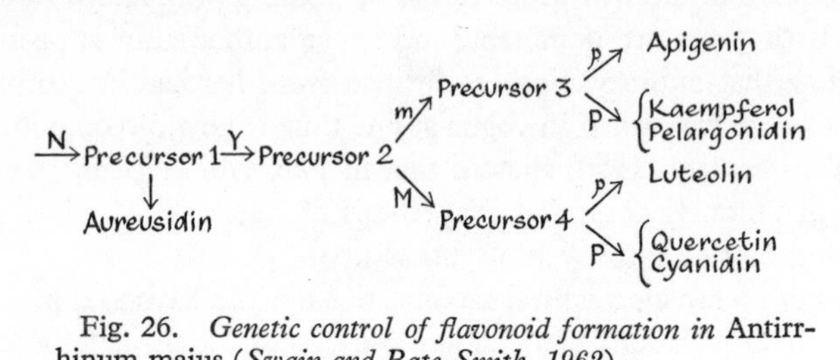

Fig. 26. *Genetic control of flavonoid formation in* Antirrhinum majus (*Swain and Bate-Smith, 1962*)

(1) N controls the formation of all the flavonoids. The recessive homozygote nn contains only cinnamic acids, while the other varieties contain aureusidin (aurone) and various flavonoids, in addition to these acids.

(2) Y controls the proportions of aureusidin in relation to other flavonoids in such a way that YY genotypes are poorer in aureusidin than yy types.

(3) P controls 3-hydroxylation. The recessive homozygotes pp contain only the flavones apigenin or luteolin, whereas the dominant homozygotes PP contain, in addition to flavone, 3-hydroxylated compounds, i.e. flavonols and anthocyanidins.

(4) M controls hydroxylation in the 3′-position of all the flavonoids e.g. producing luteolin in place of apigenin except the aurones; aureusidin (with two B-ring hydroxyls) is present in both MM and mm types and is consequently formed by an independent pathway. M must intervene before P, since dihydroxylation affects either flavone alone, or simultaneously flavone, flavonol and anthocyanidin.

8.19. *Genetic control of the formation of flavonoids in* Phaseolus vulgaris
The inheritance of pigmentation in bean seeds has been studied by Feenstra (1960), who found that two pairs of genes (V-v^{lae} and Sh-sh) and one multiallelic pair (C^r-C-C^u) controlled flavonoid synthesis:

(1) Sh-sh controls the formation of leucoanthocyanins.

(2) V-v^{lae} controls the hydroxylation of the B ring of flavonoids (2 or 3OH).

(3) The multiallelic series controls the formation of anthocyanidins and flavonols. With c^u, neither of these substances is formed. C leads to the synthesis of both types of phenolic only if there is trihydroxylation of the B ring, that is, in the presence of V; in the case of dihydroxylation (v^{lae}), only the flavonols appear. C^r has an opposite effect; it leads to dihydroxylated anthocyanidins and flavonols (in the presence of v^{lae}) or to the trihydroxylated anthocyanidins only (in the presence of V).

There is, thus, a relation between the formation of flavonols and anthocyanidins on one hand, and hydroxylation of the B ring of these flavonoids on the other. An analogous phenomenon has been observed in *V. vinifera* (P. Ribéreau-Gayon, 1964b). White grapes contain kaempferol (1OH) and quercetin (2OH); black grapes have, in addition, myricetin (3OH), as well as mainly trihydroxylated anthocyanidins.

8.20. *Application of anthocyanin genetics to wine analysis*

The so-called European varieties of grape vine, which traditionally provide grapes destined for consumption and for transformation into wine, are all derived from *V. vinifera*; however, the culture of hybrid varieties (resulting from various crosses between *V. vinifera* and other species of the genus *Vitis*) has spread recently, their production being cheaper, but the quality of their wines not reaching the high level of which *V. vinifera* is capable. There is, thus, a need for a chemical method for distinguishing whether a particular wine is derived from *V. vinifera* or is of hybrid origin; it is evident that tasting is not sufficiently objective to make this distinction, particularly where hybrid and *V. vinifera* wines have been blended. The study of grape anthocyanins has provided such a method for classifying red wines (P. Ribéreau-Gayon, 1953, 1960, 1963).

This method is based on the following observations:

(1) Anthocyanins in the form of diglucosides are not found in the fruit of *V. vinifera*.

(2) The presence of anthocyanin diglucosides in the fruit is characteristic of *V. riparia* and *V. rupestris*, two species most frequently used in hybridisation.

(3) The 'presence of diglucosides' is transmitted in the course of hybridisation as a dominant character. The observations of

Boubals *et al.* (1962) and P. Ribéreau-Gayon (1964a) show that diglucoside synthesis (from monoglucoside) is controlled by only one gene. For example, 32 crossing between a Dd heterozygote and a dd recessive homozygote gave 17 offspring possessing diglucosides and 15 without diglucosides; this approximates closely to the predicted 1 : 1 ratio. As a result, the majority of hybrids in production possess anthocyanin diglucosides and only a small proportion of them have the monoglucosides as in *V. vinifera.*

The method therefore consists of the identification of anthocyanin diglucosides: their presence proves the hybrid nature of a grape or wine but their absence is not sufficient to show that the grape or wine is from *V. vinifera* alone. Diglucoside is identified by paper chromatography using a solution of citric acid (6 g/litre) as solvent (G. Ribéreau-Gayon and E. Peynaud, 1958). This technique (see Fig. 27) gives either one

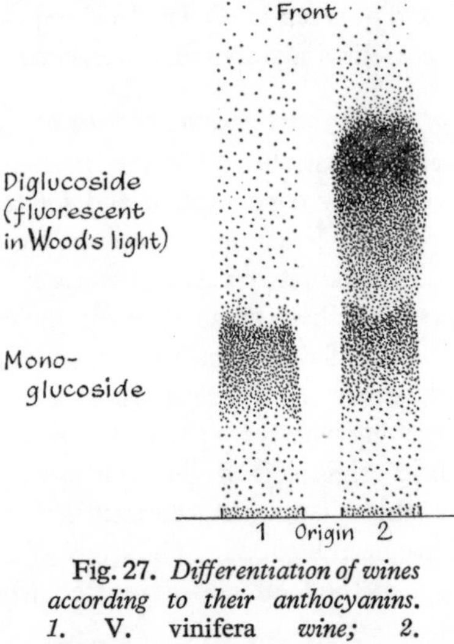

Fig. 27. *Differentiation of wines according to their anthocyanins. 1.* V. vinifera *wine; 2. majority of wines of hybrid origin (P. Ribéreau-Gayon, 1963).*

spot (monoglucoside) in the case of *V. vinifera*, or two spots (monoglucoside and diglucoside) for hybrids. The presence of the second spot, corresponding to diglucoside, can be confirmed by its brick-red fluorescence in Wood's light.

This method is of considerable economic importance and it is now widely used in the control of high grade French wines which must not contain hybrids. In Germany, legislation forbids the production and sale of hybrid wines and so all wines exported from France have to undergo prior analysis by chromatography. This method has been officially recognised by French legislation (1963) and is recognised by tribunals.

REFERENCES

ABE, Y. & HAYASHI, K. 1956. *Bot. Mag. Tokyo*, **69**, 577.

ALBACH, R. F. & REDMAN, G. H. 1969. *Phytochem.* **8**, 127.

ALBACH, R. F., WEBB, A. D. & KEPNER, R. E. 1965. *J. Food Sci.* **30**, 620.

ALIBERT, G., MARIGO, G. & BOUDET, A. 1968. *C.r. hebd. Séanc. Acad. Sci., Paris* **267**, Séries D, 2144.

ALSTON, R. E. 1964. In: *Biochemistry of phenolic compounds*, ed. by J. B. Harborne. Academic Press, New York.

ALSTON, R. E., RÖSLER, H., NAIFEH, K. & MABRY, T. J. 1965. *Proc. natn. Acad. Sci. U.S.A.* **54**, 1458.

ALSTON, R. E. & TURNER, B. L. 1963. *Biochemical systematics.* Prentice Hall Inc, Englewood Cliffs, N.J.

ANDERSEN, R. A. & SOWERS, J. A. 1968. *Phytochem.* **7**, 293.

A.O.A.C. 1960. *Official methods of the association of official agricultural chemists*, 9th edition, Washington 4, D.C.

ARAKAWA, H. & NAKAZAKI M. 1960. *Chemy. Ind.* 73.

ARNAUD, P. 1964. *Cours de chimie organique.* Gauthier-Villars, Paris.

ARTHUR, H. R., WUI, W. H. & MA, C. N. 1956. *J. chem. Soc.* 632.

ASEN, S. & BUDIN, P. S. 1966. *Phytochem.* **5**, 1257.

ASEN, S. & JURD, L. 1967. *Phytochem.* **6**, 577.

AULIN-ERDTMAN, G. 1957. In: *Les hétérocycles oxygénés.* Colloquium of C.N.R.S., Paris.

BAKER, W. & OLLIS, W. D. 1961. In: *Recent developments in the chemistry of natural phenolic compounds*, ed. by W. D. Ollis. Pergamon Press, Oxford.

BARCHEWITZ, P. 1953. *C.r. hebd. Séanc. Acad. Sci., Paris*, **237**, 237.

BARNES, H. M., FELDMAN, J. R. & WHITE, W. V. 1950. *J. Am. chem. Soc.* **72**, 4178.

BARNOUD, F. 1965. *Bull. Soc. Physio. végé.* **11**, 35.

BATE-SMITH, E. C. 1948. *Nature*, **161**, 835.

BATE-SMITH, E. C. 1949. *Biochem. Soc. Symp.* (Cambridge), **3**, 52.
BATE-SMITH, E. C. 1953. *J. exp. Bot.* **4**, 1.
BATE-SMITH, E. C. 1954a. *Advances in food research*, Vol. V. Academic Press New York.
BATE-SMITH, E. C. 1954b. *Biochem. J.* **58**, 122.
BATE-SMITH, E. C. 1954*c*. *Chemy. Ind.*, 1457.
BATE-SMITH, E. C. 1954*d*. *Food*, **23**, 124.
BATE-SMITH, E. C. 1956a. *Chemy. Ind.* 32.
BATE-SMITH, E. C. 1956b. *Scient. Proc. R. Dubl. Soc.* **27**, 365.
BATE-SMITH, E. C. 1958. *Proc. Linn. Soc. Lond.* **169**, 198.
BATE-SMITH, E. C. 1962a. In: *Wood extractives*, ed. by W. E. Hillis. Academic Press, New York.
BATE-SMITH, E. C. 1962b. *J. Linn. Soc.* (*Bot.*) **58**, 95.
BATE-SMITH, E. C. 1964. In: *Methods in polyphenol chemistry*, ed. by J. B. Pridham. Pergamon Press, Oxford.
BATE-SMITH, E. C. 1965. *Bull. Soc. bot. Fr.* (Mémoires) 16.
BATE-SMITH, E. C. 1968. *J. Linn. Soc.* (*Bot.*) **60**, 325.
BATE-SMITH, E. C., DAVENPORT, S. M. & HARBORNE, J. B. 1967. *Phytochem.* **6**, 1407.
BATE-SMITH, E. C. & LERNER, N. H. 1954. *Biochem. J.* **58**, 126.
BATE-SMITH, E. C. & METCALFE, C. R. 1957. *J. Linn. Soc.* (*Bot.*) **55**, 362.
BATE-SMITH, E. C. & RIBÉREAU-GAYON, P. 1958. *Qualitas Pl. Mater. veg.* **5**, 189.
BATE-SMITH, E. C., SELL, P. D. & WEST, C. 1968. *Phytochem.* **7**, 1165.
BATE-SMITH, E. C. & SWAIN, T. 1953a. *J. chem. Soc.* **450**, 2185.
BATE-SMITH, E. C. & SWAIN, T. 1953b. *Chemy. Ind.* 277.
BATE-SMITH, E. C. & SWAIN, T. 1956. In: *The chemistry of vegetable tannins.* Society of Leather Trades' Chemists, Croydon (England).
BATE-SMITH, E. C. & SWAIN, T. 1967. *Nature* **213**, 1033.
BATE-SMITH, E. C. & WESTALL, R. G. 1950. *Biochim. biophys. acta* **4**, 427.
BAYER, E. 1958. *Chem. Ber.* **91**, 1115.
BAYER, E. 1966. *Angew. Chem.* **78**, 834.
BAYER, E., NEHTER, K. & EGETER, H. 1960. *Chem. Ber.* **93**, 2871.
BEALE, G. H. 1941. *J. Genet.* **42**, 197.
BELLAMY, L. J. 1956. *The infrared spectra of complex molecules.* Methuen, London.
BENDZ, G., MARTENSSON, O. & TERENIUS, L. 1962. *Acta chem. scand.* **16**, 1183.
BHATIA, V. K. & SESHADRI, T. R. 1967. *Phytochem.* **6**, 1033.
BILKEY, G. 1966. In: *Biosynthesis of aromatic compounds.* Pergamon Press, Oxford.

BILLET, D. 1966. In: *Actualités de phytochimie fondamentale*, 2nd series ed. by C. Mentzer. Masson, Paris.

BILLET, D., MASSICOT, J., MERCIER, C., ANKER, D., MATSCHENKO, A., MENTZER, C., CHAIGNEAU, M., VALDNER, G. & PACHÉCO, H. 1965. *Bull. Soc. chim.* 3006.

BIRCH, A. J. 1962. In: *Chemistry of flavonoid compounds*, ed. by T. A. Geissman. Pergamon Press, Oxford.

BIRCH, A. J. 1963. In: *Chemical plant taxonomy*, ed. by T. Swain. Academic Press, New York.

BIRCH, A. J. & DONOVAN, F. W. 1953. *Aust. J. chem.* **6**, 360.

BIRKOFER, L. 1965. *Planta med. Allm.* **13**, 445.

BLANK, F. 1958. In: *Encyclopedia of plant physiology*, Vol. X, ed. by W. Ruhland. Springer Verlag, Berlin.

BOHM, B. A. 1965. *Chem. Rev.* **65**, 447.

BOTTOMLEY, W., SMITH, H. & GALSTON, A. W. 1966. *Phytochem.* **5**, 117.

BOUBALS, D., CORDONNIER, R. & PISTRE, R. 1962. *C.r. hebd. Séanc. Acad. Agric. Fr.* **48**, 201.

BOUDET, A. & COLONNA, J. P. 1968. *C.r. hebd. Séanc. Acad. Sci., Paris* **226**, series D, 2256.

BRADFIELD, A. E. & BATE-SMITH, E. C. 1950. *Biochim. biophys. acta* **4**, 441.

BRADFIELD, A. E. & FLOOD, A. E. 1952. *J. chem. Soc.*, 4470.

BRIGGS, L. H. & COLEBROOK, L. D. 1962. *Spectrochim. Acta* **18**, 939.

BROWN, S. A. 1964. In: *Biochemistry of phenolic compounds*, ed. by J. B. Harborne. Academic Press, New York.

BU'LOCK, T. D. 1965. *The Biosynthesis of natural products.* McGraw Hill, London.

BURGES, N. A. 1963. In: *Enzyme chemistry of phenolic compounds*, ed. by J. B. Pridham. Pergamon Press, Oxford.

BUTLER, W. L., HENDRICKS, S. B. & SIEGELMAN, H. W. 1965. In: *Chemistry and biochemistry of plant pigments*, ed. by T. W. Goodwin. Academic Press, New York.

CADMAN, C. H. 1960. In: *Phenolics in plants in health and disease*, ed. by J. B. Pridham. Pergamon Press, Oxford.

CATROUX, G., HENNEQUIN, J. R., FOURNIER, J. C. & BLACHÈRE, H. 1967. *C.r. hebd. Séanc. Acad. Sci., Paris.* **264**, Series D, 2533.

CHALLICE, J. S. & WILLIAMS, A. H. 1968a. *Phytochem.* **7**, 119.

CHALLICE, J. S. & WILLIAMS, A. H. 1968b. *Phytochem.* **7**, 1781.

CHANDLER, B. V. 1958. *Nature*, **182**, 933.

CHANDLER, B. V. & HARPER, K. A. 1958. *Nature*, **181**, 131.

CHANDLER, B. V. & HARPER, K. A. 1961. *Aust. J. chem.* **14**, 586.

CHANDLER, B. V. & HARPER, K. A. 1962. *Aust. J. chem.* **15**, 114.

CHAPON, L., CHOLLOT, B. & URION, E. 1961. *Bull. Soc. Chim. biol.* **43**, 429.

CHARAUX, C. 1925. *Bull. Soc. Chim. biol.* **7**, 1053.

CHOPIN, J., BOUILLANT, M. L. & LEBRETON, P. 1964. *Bull. Soc. chim.* 1038.

CHOPIN, J. 1966. In: *Actualités de phytochimie fondamentale*, 2nd series, ed. by C. Mentzer. Masson, Paris.

CHOUTEAU, J. & LOCHE, J. 1962. *C.r. hebd. Séanc. Acad. Sci., Paris* **254**, 2064.

CLARK-LEWIS, J. W. 1962. In: *The chemistry of flavonoid compounds*, ed. by T. A. Geissman. Pergamon Press, Oxford.

CO, H. & MARKAKIS, P. 1966. *Phytochem.* **5**, 755.

COLTHUP, N. B. 1950. *J. opt. Soc. Am.* **40**, 397.

COMBIER, H. & LEBRETON, P. 1968. *C.r. hebd. Séanc. Acad. Sci., Paris* **267**, Series D, 421.

COMMONER, G. 1948. *Ann. Mo. bot. Gdn*, **35**, 239 (quoted by SANNIÉ & SAUVAIN, 1952).

COMTE, P., VILLE, A., ZWINGELSTEIN, G. & FAVRE-BONVIN, J. 1958a. *Bull. Soc. chim.* 1355.

COMTE, P., VILLE, A., ZWINGELSTEIN, G. & FAVRE-BONVIN, J. 1958b. *Bull. Soc. Chim. biol.* **40**, 1117.

COMTE, P., VILLE, A., ZWINGELSTEIN, G. & FAVRE-BONVIN, J. 1960. *Bull. Soc. Chim. biol.* **42**, 1079.

CONN, E. E. 1964. In: *Biochemistry of phenolic compounds*, ed. by J. B. Harborne. Academic Press, New York.

CORNER, J. J. & SWAIN, T. 1965. *Nature*, **207**, 634.

CORSE, J., LUNDIN, R. E. & WAISS, A. C. 1965. *Phytochem.* **4**, 527.

CRAM, D. J. & HAMMOND, G. S. 1963. *Chimie organique.* Les Presses de l'Université Laval, Quebec.

CREASY, L. L. 1968. *Phytochem.* **7**, 441.

CREASY, L. L. & SWAIN, T. 1965. *Nature*, **208**, 151.

CREASY, L. L. & SWAIN, T. 1966. *Phytochem.* **5**, 501.

CRUICKSHANK, J. A. M. & PERRIN, D. R. 1964. In: *Biochemistry of phenolic compounds*, ed. by J. B. Harborne. Academic Press, New York.

DAVIS, B. D. 1955. *Advances in Enzymology*, **16**, 247.

DAVIS, P. H. & HEYWOOD, V. H. 1963. *Principles of angiosperm taxonomy*. Oliver and Boyd, Edinburgh.

DAYTON, T. O. 1956. *J. Genet.* **54**, 249.

DEAN, F. M. 1963. *Naturally occurring oxygen ring compounds.* Butterworths, London.

DICKINSON, D. & GAWLER, J. H. 1956. *J. Sci. Fd Agric.* **7**, 669.

DIEMAIR, W. & POLSTER, A. 1967. *Z. Lebensmittelunters. u.-Forsch:* **134**, 345.
DOURMICHIDZE, S. V. 1950. *Dokl. (Proc.) Acad. Sci. U.S.S.R., Biochem.* **73**, 987.
DOURMICHIDZE, S. V. 1950. *Dokl. (Proc.) Acad. Sci. U.S.S.R., Biochem.* **77**, 859.
DOURMICHIDZE, S. V. 1955. *Tannins et anthocyanes des raisins et des vins.* Academy of Sciences of the U.S.S.R., Moscow.
DREWES, S. E. & ROUX, D. G. 1966. *Biochem. J.* **98**, 493.
EDMONSON, Y. H. & THIMANN, K. V. 1950. *Archs. Biochem.* **25**, 79.
EDWARDS, K. G. & STOKER, J. R. 1967. *Phytochem.* **6**, 655.
EGGER, K. 1961. *Z. analyt. Chem.* **182**, 161.
EGGER, K. 1964. *Planta med.* **12**, 265.
EGGER, K. & KEIL, M. 1965. *Z. analyt. Chem.* **210**, 201.
EIGEN, E., BLITZ, M. & GUNSBERG, E. 1957. *Archs. Biochem. Biophys.* **68**, 501.
EL BASYOUNI, S. Z., CHEN, D., IBRAHIM, R. K., NEISH, A. C. & TOWERS, G. H. N. 1964. *Phytochem.* **3**, 485.
EL BASYOUNI, S. Z. & NEISH, A. C. 1966. *Phytochem.* **5**, 683.
ENDO, T. 1957. *Nature,* **179**, 378.
ERDTMAN, H. 1957. In: *Les hétérocycles oxygénés.* Colloquium of C.N.R.S. Paris.
FAIRBAIRN, J. W. 1959. In: *The pharmacology of plant phenolics.* Academic Press, New York.
FEENSTRA, W. J. 1960. *Biochemical aspects of seedcoat colour inheritance in* Phaseolus vulgaris. H. Veenman en Zonen N. V., Wageningen.
FEENSTRA, W. J., JOHNSON, B. L., RIBÉRAU-GAYON, P. & GEISSMAN, T. A. 1963. *Phytochem.* **2**, 273.
FEENY, P. P. & BOSTOCK, H. 1968. *Phytochem.* **7**, 871.
FIESER, L. F. & FIESER, M. 1956. *Organic chemistry.* Reinhold Publishing Corporation, New York.
FINKLE, B. J. & NELSON, R. F. 1963. *Biochim. biophys. Acta,* **78**, 747.
FORSYTH, W. G. C. & QUESNEL, V. C. 1957a. *Biochem. J.* **65**, 177.
FORSYTH, W. G. C. & QUESNEL, V. C. 1957b. *Biochim. biophys. Acta,* **25**, 155.
FORSYTH, W. G. C., QUESNEL, V. C. & ROBERTS, J. B. 1960. *Biochim. biophys. Acta,* **37**, 322.
FORSYTH, W. G. C. & ROBERTS, J. B. 1960. *Biochem. J.* **74**, 374.
FREUDENBERG, K. 1959. *Nature,* **183**, 1152.
FREUDENBERG, K. & WEINGES, K. 1961. *Tetrahedron Lett.* **8**, 267.
FREUDENBERG, K. & WEINGES, K. 1962. In: *The chemistry of flavonoid compounds,* ed. by T. A. Geissman. Pergamon Press, Oxford.

Fritig, B., Hirth, L. & Ourisson, G. 1966. *C.r. hebd. Séanc. Acad. Sci., Paris*, **263**, Series D, 860.

Fuleki, T. & Francis, F. J. 1967. *J. of Chromat.* **26**, 404.

Fuleki, T. & Francis, F. J. 1968. *J. Food Sci.* **33**, 72, 78 and 266.

Gautheret, R. J. 1949. *La culture des tissus végétaux.* Masson, Paris.

Geissman, T. A. 1955. In: *Modern methods of plant analysis*, vol. III, ed. by K. Peach and V. Tracey. Springer Verlag, Berlin.

Geissman, T. A. 1962. *The chemistry of flavonoid compounds.* Pergamon Press, Oxford.

Geissman, T. A. 1963a. In: *Biogenesis of natural compounds*, ed. by P. Bernfeld. Pergamon Press, Oxford.

Geissman, T. A. 1963b. In: *Comprehensive biochemistry*, vol. 9, ed. by M. Florkin and E. H. Stotz. Elsevier, London.

Geissman, T. A. 1965. *Principes de chimie organique.* Dunod, Paris.

Geissman, T. A. & Dittmar, H. F. K. 1965. *Phytochem.* **4**, 359.

Geissman, T. A. & Harborne, J. B. 1955. *Archs. Biochem. Biophys.* **55**, 447.

Geissman, T. A. & Hinreiner, E. 1952. *Bot. Rev.* **18**, 77–164.

Geissman, T. A., Jorgensen, E. C. & Johnson, B. L. 1954. *Archs. Biochem. Biophys.* **49**, 368.

Geissman, T. A. & Yoshimura, N. N. 1966. *Tetrahedron Lett.* **24**, 2669.

Genevois, L. 1955. *Annls. Nutr. Aliment.* **9**, A. 295.

Genevois, L. 1956. *Bull. Soc. Chim. biol.* **38**, 7.

Goldstein, J. & Swain, T. 1963. *Phytochem.* **2**, 371.

Goodwin, R. H. & Pollock, B. M. 1954. *Archs. Biochem. Biophys.* **49**, 1.

Goodwin, T. W. (ed.) 1965. *Chemistry and biochemistry of plant pigments.* Academic Press, London & New York.

Govindarajan, V. S. & Mathew, A. G. 1965. *Phytochem.* **4**, 985.

Griffiths, L. A. 1959. *J. exp. Bot.* **10**, 437.

Gripenberg, J. 1962. In: *The chemistry of flavonoid compounds*, ed. by T. A. Geissman. Pergamon Press, Oxford.

Grisebach, H. 1957. *Z. Naturf.* **12** b, 227.

Grisebach, H. 1961. In: *Recent developments in the chemistry of natural phenolic compounds*, ed. by W. D. Ollis. Pergamon Press, Oxford.

Grisebach, H. 1965. In: *Chemistry and biochemistry of plant pigments*, ed. by T. W. Goodwin. Academic Press, London & New York.

Grisebach H. 1967. *Biosynthetic patterns in microorganisms and higher plants.* John Wiley and Sons, New York.

Grisebach, H. & Bopp, M. 1959. *Z. Naturf.* **14** b, 485.

Grisebach, H. & Brandner, G. 1962. *Biochim biophys. Acta.* **60**, 51.

Grouiller, A. 1966. *Bull. Soc. chim.* 2405.

HANSON, K. R. 1966. *Phytochem.* **5**, 491.

HANSON, K. R. & ZUCKER, M. 1963. *J. biol. Chem.* **238**, 1105.

HARBORNE, J. B. 1958. *Biochem. J.* **70**, 22.

HARBORNE, J. B. 1959a. *Chromatographic reviews*, Vol. 1, ed. by M. Lederer. Elsevier, Amsterdam.

HARBORNE, J. B. 1959b. *J. of Chromat.* **2**, 581.

HARBORNE, J. B. 1960a. *Biochem. J.* **74**, 262.

HARBORNE, J. B. 1960b. *Biochem. J.* **74**, 270.

HARBORNE, J. B. 1962a. In: *The chemistry of flavonoid compounds*, ed by T. A. Geissman. Pergamon Press, Oxford.

HARBORNE, J. B. 1962b. *Fortschr. Chem. org. Nat. Staffe.* **20**, 165.

HARBORNE, J. B. 1963. *Phytochem.* **2**, 85.

HARBORNE, J. B. 1964a. *Phytochem.* **3**, 151.

HARBORNE, J. B. 1964b. In: *Biochemistry of phenolic compounds*, ed. by J. B. Harborne. Academic Press, New York.

HARBORNE, J. B. (ed.) 1964c. *Biochemistry of phenolic compounds.* Academic Press, New York.

HARBORNE, J. B. 1964d. In: *Methods in polyphenol chemistry*, ed. by J. B. Pridham. Pergamon Press, Oxford.

HARBORNE, J. B. 1965a. *Phytochem.* **4**, 107.

HARBORNE, J. B. 1965b. In: *Chemistry and biochemistry of plant pigments*, ed. by T. W. Goodwin. Academic Press, New York.

HARBORNE, J. B. 1965c. *Phytochem.* **4**, 647.

HARBORNE, J. B. 1966a. *Phytochem.* **5**, 111.

HARBORNE, J. B. 1966b. *Phytochem.* **5**, 589.

HARBORNE, J. B. 1967a. *Comparative biochemistry of the flavonoids.* Academic Press, New York.

HARBORNE, J. B. 1967b. *Phytochem.* **6**, 1415.

HARBORNE, J. B. 1968. *Phytochem.* **7**, 1215.

HARBORNE, J. B. 1969a. *Phytochem.* **8**, 177.

HARBORNE, J. B. 1969b. *Phytochem.* **8**, 419.

HARBORNE, J. B. & CORNER, J. J. 1960. *Biochem. J.* **76**, 53 P.

HARBORNE, J. B. & CORNER, J. J. 1961. *Biochem. J.* **81**, 242.

HARBORNE, J. B. & GEISSMAN, T. A. 1956. *J. Am. chem. Soc.* **78**, 829.

HARBORNE, J. B. & SHERRATT, H. S. A. 1957. *Biochem. J.* **65**, 23 P.

HARBORNE, J. B. & SHERRATT, H. S. A. 1961. *Biochem. J.* **78**, 298.

HARBORNE, J. B. & SIMMONDS, N. W. 1964. In: *Biochemistry of phenolic compounds*, ed. by J. B. Harborne. Academic Press, New York.

HARBORNE, J. B. & SWAIN, T. (eds.) 1969. *Perspectives in phytochemistry.* Academic Press, New York.

HARKISS, K. J. 1965. *Nature*, **205**, 78.

HARPER, K. A. & CHANDLER, B. V. 1967a. *Aust. J. Chem.* **20**, 731.

HARPER, K. A. & CHANDLER, B. V. 1967b. *Aust. J. Chem.* **20**, 745.
HARRIS, G. & RICKETTS, R. W. 1959. *7th European Brewery Conv., Proc. Cong.*, Rome.
HASLAM, E. 1966. *Chemistry of vegetable tannins.* Academic Press, New York.
HASLAM, E. 1967. *J. chem. Soc.* **18**, 1734.
HATHWAY, D. E. 1962. In: *Wood extractives*, ed. by W. E. Hillis. Academic Press, New York.
HAVELANGE, A. & SCHUMACKER, R. 1966. *Bull. Soc. Sci. r. Liège*, **35**, 125.
HAYASHI, K. 1962. In: *The chemistry of flavonoid compounds*, ed. by T. A. Geissman. Pergamon Press, Oxford.
HAYASHI, K. & ISAKA, K. 1946. *Proc. Japan Acad.* **22**, 256.
HAYASHI, K., SAITÔ, N. & MITSUI, S. 1961. *Proc. Japan Acad.* **37**, 393.
HELLER, R. 1948a. *C.r. Séanc. Soc. Biol.* **142**, 768.
HELLER, R. 1948b. *C.r. Séanc. Soc. Biol.* **142**, 947.
HENDRICKS, S. B. & BORTHWICK, H. A. 1965. In: *Chemistry and Biochemistry of plant pigments*, ed. by T. W. Goodwin. Academic Press, New York.
HENGLEIN, F. A. & KRAMER, J. 1959. *Chem. Ber.* **92**, 2585.
HENRY, L. & MOLHO, D. 1957. In: *Les hétérocycles oxygénés.* Colloquium of C.N.R.S., Paris.
HERGERT, H. L. 1962. In: *The chemistry of flavonoid compounds*, ed. by T. A. Geissman. Pergamon Press, Oxford.
HERGERT, H. L. & KURTH, E. F. 1953. *J. Am. chem. Soc.* **75**, 1962.
HERMANN, K. 1959. *Mitt. Geb. Lebensmittelunters. u. Hyg.* **50**, 121.
HERMANN, K. 1963. *Weinberg und Keller*, **10**, 154 and 208.
HERMANN, K. 1965. *Handbuch der Lebensmittelchemie.* Berlin.
HEYWOOD, V. H. 1966. In: *Comparative phytochemistry*, ed. by T. Swain. Academic Press, London & New York.
HEYWOOD, V. H. (ed.) 1968. *Modern methods in plant taxonomy.* Academic Press, London & New York.
HILLIS, W. E. (ed.) 1962. *Wood extractives.* Academic Press, New York.
HILLIS, W. E. 1966. *Phytochem.* **5**, 541.
HILLIS, W. E. 1967a. *Phytochem.* **6**, 259.
HILLIS, W. E. 1967b. *Phytochem.* **6**, 275.
HILLIS, W. E. 1967*c*. *Phytochem.* **6**, 373.
HILLIS, W. E. 1967*d*. *Phytochem.* **6**, 845.
HILLIS, W. E. & CARLE, A. 1960. *Aust. J. Chem.* **13**, 390.
HILLIS, W. E. & HORN, D. H. S. 1965. *Aust. J. Chem.* **13**, 531.
HILLIS, W. E. & SWAIN, T. 1959. *J. Sci. Fd Agric.* **2**, 135.
HÖRHAMMER, L., STICH, L. & WAGNER, H. 1959. *Naturwissenschaften*, **48**, 358.

Hörhammer, L. & Wagner, H. 1961. In: *Recent developments in the chemistry of natural phenolic compounds*, ed. by W. D. Ollis. Pergamon Press.
Horowitz, R. M. 1964. In: *Biochemistry of phenolic compounds*, ed. by J. B. Harborne. Academic Press, New York.
Huang, H. T. 1955. *J. agric. Fd Chem.* **3**, 141.
Hurst, H. M. & Harborne, J. B. 1967. *Phytochem.* **6**, 1111.
Ibrahim, R. K. & Towers, G. H. N. 1960. *Archs. Biochem. Biophys.* **87**, 125.
Ibrahim, R. K., Towers, G. H. N. & Gibbs, R. D. 1962. *J. Linn. Soc. (Bot.)*, **58**, 223.
Ice, C. H. & Wender, S. H. 1952. *Analyt. Chem.* **24**, 1616.
Imaseki, H., Wheeler, R. E. & Geissman, T. A. 1965. *Tetrahedron Lett.* **23**, 1785.
Inglett, G. E. 1958. *J. org. Chem.* **23**, 93.
Ishikura, N. & Hayashi, K. 1966. *Bot. Mag., Tokyo*, **79**, 156.
Jacquin, F. 1963. *Contribution à l'étude des processus de formation et d'évolution de divers composés humiques.* Natural science thesis, Nancy.
Jan, J. & Reid, W. W. 1959. *Chemy Ind.* 655.
Jones, R. M. & Sandorfy. 1956. *Chemical applications of spectroscopy.* Interscience Publishers, London.
Joslyn, M. A. & Dittmar, H. F. K. 1967. *Am. J. Enol. Vitic.* **18**, 1.
Joslyn, M. A. & Goldstein, J. L. 1964a. *Science*, **143**, 954.
Joslyn, M. A. & Goldstein, J. L. 1964b. *Advances in food research*, vol. 13. Academic Press, New York.
Joslyn, M. A. & Goldstein, J. L. 1964c. *Agric. Food Chem.* **12**, 511.
Jurd, L. 1956. *Archs. Biochem. Biophys.* **63**, 376.
Jurd, L. 1962a. In: *The chemistry of flavonoid compounds*, ed. by T. A. Geissman. Pergamon Press, New York.
Jurd, L. 1962b. In: *Wood extractives*, ed. by W. E. Hillis. Academic Press, New York.
Jurd, L. 1963. *J. org. Chem.* **28**, 987.
Jurd, L. 1964a. *J. Food Sci.* **29**, 16.
Jurd, L. 1964b. *J. org. Chem.* **29**, 2602.
Jurd, L. 1965. *Tetrahedron Lett.* **21**, 3707.
Jurd, L. 1966. *Chemy. Ind.* 1683.
Jurd, L. 1967. *Tetrahedron Lett.* **23**, 1057.
Jurd, L. 1968a. *Experientia*, **24**, 858.
Jurd, L. 1968b. *Tetrahedron Lett.* **24**, 2801.
Jurd, L. 1968c. *Tetrahedron Lett.* **23**, 4449.
Jurd, L. & Asen, S. 1966. *Phytochem.* **5**, 1263.
Jurd, L. & Bergot, B. J. 1965. *Tetrahedron Lett.* **21**, 3697.
Jurd, L. & Geissman, T. A. 1956. *J. org. Chem.* **21**, 1395.

Jurd, L. & Harborne, J. B. 1968. *Phytochem.* **7**, 1209.
Jurd, L. & Horowitz, R. M. 1957. *J. org. Chem.* **22**, 1618.
Jurd, L. & Lundin, R. 1968. *Tetrahedron Lett.* **24**, 2653.
Jurd, L. & Waiss, A. C. 1964. *Chemy. Ind.* 1708.
Kahnt, G. 1967. *Phytochem.* **6**, 755.
Kameswaramma, A. & Seshadri, T. R. 1947. *Proc. Indian Acad. Sci., A*, **25**, 43.
Karrer, P. 1928. *Bull. Soc. chim.* **43**, 1041.
Karrer, P. 1945. In: *Traité de chimie organique*, vol. XVIII, ed. by V. Grignard, Masson, Paris.
Karrer, P. & de Meuron, G. 1932. *Helv. chim. Acta*, **15**, 507 and 1212.
Karrer, P. & Strong, F. M. 1936. *Helv. chim. Acta*, **19**, 25.
Karrer, P. & Weber, H. M. 1936. *Helv. chim. Acta*, **19**, 1025.
Keith, E. S. & Powers, J. J. 1965. *J. Agric. Fd Chem.* **13**, 577.
Khanna, S. K., Viswanathan, P. N., Krishann, P. S. & Sannal, G. C. 1968. *Phytochem.* **7**, 1513.
King, H. G. C. & White, T. 1956. In: *The chemistry of vegetable tannins.* Society of Leather Trades' Chemists, Croydon, England.
Kuhn, H. & Sperling, W. 1960. *Experienta*, **16**, 237.
Kursanov, A. L. & Zaprometov, M. N. 1949. *Biokhimiya*, **14**, 467.
Kursanov, A. L. & Zaprometov, M. N. 1952. *Atompraxis*, **4**, 280, (cited by Neish, 1964).
Laborde, J. 1908. *C.r. hebd. Séanc. Acad. Sci., Paris*, June and December.
Laborde, J. 1910. *Revue de Viticulture*, **33**, 206.
Lavollay, J., Legrand, G., Lehongre, G. & Neumann, J. 1958. *Qualitas Pl. Mater. veg.*, **3/4**, 508.
Lavollay, J. & Neumann, J. 1959. In: *The pharmacology of plant phenolics*, ed. by J. W. Fairbairn. Academic Press, New York.
Lavollay, J. & Vignau, M. 1943. *C.r. hebd. Séanc. Acad. Sci., Paris*, **217**, 87.
Lawrence, W. J. C. & Price, J. R. 1940. *Biol. Rev.* **15**, 35.
Lawrence, W. J. C., Price, J. R., Robinson, G. M. & Robinson, R. 1939. *Trans. R. Soc.* (London) (B), **230**, 149.
Lebreton, P. 1962. *Contribution à l'étude des flavonoides chez* Humulus lupulus *L. et autres Urticales.* Thèse Sciences naturelles, Lyon.
Lebreton, P. 1964. *Bull. Soc. bot. Fr.* **111**, 80.
Lebreton, P., Jay, M. & Voirin, B. 1967. *Chim. analyt.* **49**, 375.
Lebreton, P. & Meneret, G. 1964. *Bull. Soc. bot. Fr.* **111**, 69.
Lederer, E. 1959, 1960. *Chromatographie en chimie organique et biologique*, Vols. I, II. Masson, Paris.
Lederer, E. & Lederer, M. 1957. *Chromatography*, 2nd edition. Elsevier, Amsterdam.

LEVY, C. C. & ZUCKER, M. 1960. *J. biol. Chem.* **236**, 2043.

LEWAK, S. 1968. *Phytochem.* **7**, 665.

LI, K. G. & WAGENKNECHT, A. G. 1956. *J. Am. chem. Soc.* **78**, 979.

LOCHE, J. & CHOUTEAU, J. 1963. *C.r. hebd. Séanc. Acad. Agric. Fr.* **44**, 1017.

LOOMIS, W. D. & BATAILLE, J. 1966. *Phytochem.* **5**, 423.

MABRY, T. J. 1966. In: *Comparative phytochemistry*, ed. by T. E. Swain. Academic Press, New York.

MABRY, T. J., KAGAN, J. & RÖSLER, H. 1965. *Phytochem.* **4**, 177.

MABRY, T. J., MARKHAM, K. R. & THOMAS, M. B. 1969. *The systematic identification of flavonoids.* Springer Verlag Inc., New York.

MABRY, T. J., TAYLOR, A. & TURNER, B. L. 1963. *Phytochem.* **2**, 61.

MAHESH, V. B. & SESHADRI, T. R. 1955. *J. chem. Soc.* 2533.

MARKHAM, K. R. & MABRY, T. J. 1968a. *Phytochem.* **7**, 791.

MARKHAM, K. R. & MABRY, T. J. 1968b. *Phytochem.* **7**, 1197.

MARKHAM, K. R., MABRY, T. J. & SWIFT, T. W. 1968. *Phytochem.* **7**, 803.

MARKHAM, K. R., MABRY, T. J. & VOIRIN, B. 1969. *Phytochem.* **8**, 469.

MARTIN, C. 1958. *Etude de quelques déviations du métabolisme chez les plantes atteintes de maladies à virus.* Natural science thesis, Paris.

MASQUELIER, J. 1959. In: *The pharmacology of plant phenolics*, ed. by J. W. Fairbairn. Academic Press, New York.

MASQUELIER, J. 1960. Personal communication.

MASQUELIER, J. & POINT, G. 1956. *Bull. Trav. Soc. Pharm. Bordeaux*, **95**, 6.

MASQUELIER, J. & RICCI, R. 1964. *Qualitas Pl. Mater. veg.*, **11**, 244.

MASQUELIER, J., VITTE, G. & ORTÉGA, M. 1959. *Bull. Trav. Soc. Pharm. Bordeaux*, **98**, 145.

MASSICOT, J. & MARTHE, J. P. 1962. *Bull. Soc. chim.* 1962.

MASSICOT, J., MARTHE, J. P. & HEITZ, J. 1963. *Bull. Soc. chim.* 2712.

MAURICE, A. & MENTZER, C. 1954. *Bull. Soc. Chim. biol.* **36**, 369.

MAYER, W., GOLL, L., MORITZ VON ARNDT, E. & MANNSCHRECK, A. 1966. *Tetrahedron Lett.* 429.

MCFARLANE, W. D. & THOMPSON, K. D. 1964. *J. Inst. Brew.* **70**, 497.

MCFARLANE, W. D., WYE, E. & GRANT, H. L. 1955. *5th European Brewery Conv., Proc. Cong.* Baden-Baden.

MENTZER, C. 1960. *La théorie biogénétique et son application au classement des substances organiques d'origine végétale.* Editions du Muséum, Paris.

MENTZER, C. (ed.) 1964. *Actualités de phytochimie fondamentale*, 1st series. Masson, Paris.

MENTZER, C. (ed.) 1966. *Actualités de phytochimie fondamentale*, 2nd series. Masson, Paris.

MENTZER, C. & JOUANNETEAU, J. 1955. *Bull. Soc. Chim. biol.* **37**, 887.
METCHE, M. 1967. *Brass. fr.* **242**, 69.
METCHE, M., JACQUIN, F., NGUYEN, H. & URION, E. 1962. *Bull. Soc. chim.* **10**, 1763.
METCHE, M. & URION, E. 1961. *C.r. hebd. Séanc. Acad. Sci., Paris,* **252**, 356.
MICHAUD, J. & MASQUELIER, J. 1968. *Bull. Soc. Chim. biol.* **50**, 1346.
MINALE, L., PIATTELLI, & DE STEFANO, S. 1967. *Phytochem.* **6**, 703.
MINALE, L., PIATELLI, M., DE STEFANO, S. & NICOLAUS, R. A. 1966. *Phytochem.* **5**, 1037.
MIURA, H., KIHARA, T. & KAWANO, N. 1968. *Tetrahedron Lett.* **19**, 2339.
MONTIES, B. 1966. *Annls. Physiol vég. Paris* (I.N.R.A.), **8**, 49 et 101.
MOUSTAFA, E. & WONG, E. 1967. *Phytochem.* **6**, 625.
NEGRUL, A. M. 1946. *Ampelography of the USSR*, Vol. I, Ministry of Agriculture, Moscow (translation).
NEISH, A. C. 1964. In: *Biochemistry of phenolic compounds*, ed. by J. B. Harborne. Academic Press, New York.
NÉTIEN, G. & LACHARME, J. 1955. *Bull. Soc. Chim. biol.* **37**, 643.
NGUYEN, Q. H., METCHE, M. & URION, E. 1965. *C.r. hebd. Séanc. Acad. Sci., Paris,* **260**, 7047.
NORDSTRÖM, C. G. 1956. *Acta chem. scand.* **10**, 1491.
NORDSTRÖM, C. G. & SWAIN, T. 1953. *J. chem. Soc.* 2764.
NOTT, P. E. & ROBERTS, J. C. 1967. *Phytochem.* **6**, 741.
NYBOM, N. 1964. *Physiologia Pl.* **17**, 157.
OLLIS, W. D. 1962. In: *The chemistry of flavonoid compounds*, ed. by T. A. Geissman. Pergamon Press, Oxford.
OSAWA, Y. & SAITÔ, N. 1968. *Phytochem.* **7**, 1189.
PACHÉCO, H. 1956. *Bull. Soc. chim.* 1600.
PACHÉCO, H. 1957. *Bull. Soc. Chim. biol.* **39**, 971.
PACHÉCO, H. 1966. In: *Actualités de phytochimie fondamentale*, 2nd series, ed. by C. Mentzer. Masson, Paris.
PACHÉCO, H. & GROUILLER, A. 1965. *Bull. Soc. chim.* **3**, 779 and 2937.
PACHÉCO, H. & GROUILLER, A. 1966. *Bull. Soc. chim.* **10**, 3212.
PACHÉCO, H., PLA, J. & VILLE, A. 1965. *C.r. hebd. Séanc. Acad. Sci., Paris* **262**, Series D, 926.
PARIS, R. R. 1961. *Journées internationales de séparation immédiate et de chromatographie.* (G.A.M.S.) Paris.
PARIS, R. R. 1963. *J. Pharm. Belg.* 11.
PARIS, R. R. 1965. *Bull. Soc. chim.* 2597.
PARIS, R. R. 1968. *Qualitas. Pl. Mater. veg.* **16**, 244.
PARTRIDGE, S. M. 1947. *Biochem. J.* **42**, 238.

PATSCHKE, L. & GRISEBACH, H. 1968. *Phytochem.* **7**, 235.

PAUPARDIN, C. 1965. *C.r. hebd. Séanc. Acad. Sci., Paris*, **261**, Séries D, 4206.

PAUPARDIN, C. 1967. *C.r. hebd. séanc. Acad. Sci., Paris*, **264**, séries D, 2103.

PAYEN, A. 1846. *C.r. hebd. Séanc. Acad. Sci., Paris*, **23**, Séries D, 244.

PEARL, I. A. 1958. *Tappi*, **41**, 621.

PEARL, I. A. 1967. *The chemistry of lignin.* Marcel Dekker Inc., New York.

PERI, C. 1967. *Am. J. Enol. Vitic.* **1J**, 168.

PIATELLI, M., MINALE, L. & PROTA, G. 1965. *Phytochem.* **4**, 121.

PICTET, C. & BRANDENBERGER, H. 1960. *J. Chromat.* **4**, 396.

PIGMAN, W., ANDERSON, E., FISCHER, R., BUCHANAN, M. A. & BROWNING, B. L. 1953. *Tappi*, **36**, 4.

PLA, J., VILLE, A. & PACHÉCO, H. 1967. *Bull. Soc. Chim. biol.* **49**, 395.

POURRAT, H., TRONCHE, P. & POURRAT, A. 1966. *Bull. Soc. chim.* **6**, 1918.

PRIDHAM, J. B. 1956. *Analyt. Chem.* **2J**, 1967.

PRIDHAM, J. B. (ed.) 1963. *Enzyme chemistry of phenolic compounds.* Pergamon Press, Oxford.

QUESNEL, V. C. 1968. *Phytochem.* **7**, 1583.

RAMWELL, P. W., SHERRATT, H. S. A. & LEONARD, B. E. 1964. In: *Biochemistry of phenolic compounds*, ed. by J. B. Harborne. Academic Press, New York.

RANDERATH, K. 1964. *Chromatographie sur couches minces.* Gauthier-Villars, Paris.

RIBÉREAU-GAYON, J. & GARDRAT, J. 1956. *C.r. hebd. Séanc. Acad. Sci., Paris*, **234**, Séries D, 788.

RIBÉREAU-GAYON, J. & PEYNAUD, E. 1958. *Analyse et contrôle des vins.* Beranger, Paris.

RIBÉREAU-GAYON, P. 1953. *C.r. hebd. Séanc. Acad. Agric. Fr.* **39**, 800.

RIBÉREAU-GAYON, P. 1959. *Recherches sur les anthocyannes des végétaux, application au genre* Vitis. Libraire générale de l'Enseignement, Paris.

RIBÉREAU-GAYON, P. 1960a. *Dt. LebensmittRdsch.* **8**, 217.

RIBÉREAU-GAYON, P. 1960b. *C.r. hebd. Séanc. Acad. Sci. Paris*, **250**, Séries D, 591.

RIBÉREAU-GAYON, P. 1963. *Inds agric. aliment.* **80**, 1079.

RIBÉREAU-GAYON, P. 1964a. *Les composés phénoliques du raisin et du vin.* Institut national de la Recherche agronomique, Paris.

RIBÉREAU-GAYON, P. 1964b. *C.r. hebd. Séanc. Acad. Sci., Paris*, **258**, Séries D, 1335.

RIBÉREAU-GAYON, P. 1965. *C.r. hebd. Séanc. Acad. Sci., Paris*, **260**, Séries D, 341.

RIBÉREAU-GAYON, P. & JOSIEN, M. L. 1960. *Bull. Soc. chim.* 934.
RIBÉREAU-GAYON, P. & SAPIS, J.-C. 1965. *C.r. hebd. Séanc. Acad. Sci., Paris,* **261**, Séries D, 1915.
RIBÉREAU-GAYON, P. & STONESTREET, E. 1965a. *Bull. Soc. chim.* 2649.
RIBÉREAU-GAYON, P. & STONESTREET, F. 1965b. *Chim. analyt.* **48**, 188.
ROBERTS, E. A. H. 1956. In: *The chemistry of vegetable tannins.* Society of Leather Trades' Chemists, Croydon (England).
ROBERTS, E. A. H., CARTWRIGHT, R. A. & WOOD, D. J. 1956. *J. Sci. Fd. Agric.* **7**, 637.
ROBINSON, R. 1936. *Nature,* **137**, 1172.
ROBINSON, R. & ALLAN, J. 1924. *J. chem. Soc.* **125**, 2192.
ROBINSON, G. M. & ROBINSON, R. 1931. *Biochem. J.* **25**, 1687.
ROBINSON, G. M. & ROBINSON, R. 1932. *Biochem. J.* **26**, 1647.
ROBINSON, G. M. & ROBINSON, R. 1933. *Biochem. J.* **27**, 206.
ROBINSON, R. & SHINODA. 1925. *J. Am. chem. Soc.* **127**, 1979.
ROLLMAN, B., VANCRAENENBROECK, R. & LONTIE, R. 1966. *Archs int. Physiol. Biochim.* **74**, 714.
ROSENHEIM, O. 1920. *Biochem. J.* **14**, 178.
ROUBAIX DE, J. & LAZAR, O. 1960. In: *Phenolics in plants in health and disease,* ed. by J. B. Pridham. Pergamon Press, Oxford.
ROUX, D. G. 1957. *Nature,* **180**, 973.
ROUX, D. G. 1958a. *Biochem. J.* **69**, 530.
ROUX, D. G. 1958b. *J. Am. Leath. Chem. Ass.* **53**, 384.
ROUX, D. G. 1959. *Nature,* **183**, 1168.
ROUX, D. G. & EVELYN, S. R. 1958a. *J. of Chromat.* **1**, 537.
ROUX, D. G. & EVELYN, S. R. 1958b. *Biochem. J.* **69**, 530.
ROUX, D. G. & MAIHS, E. A. 1960. *J. of Chromat.* **4**, 65.
ROUX, D. G. & PAULUS, K. 1962. *Biochem. J.* **82**, 320.
ROYALS, E. E. 1964. *Advanced organic chemistry.* Prentice-Hall Inc. Englewood Cliffs, N.J.
RUSSELL, D. W. & GALSTON, A. W. 1967. *Phytochem.* **6**, 791.
SAITÔ, N. 1967. *Phytochem.* **6**, 1013.
SAITÔ, N., MITSUI & HAYASHI, K. 1961. *Proc. Japan Acad.* **37**, 485.
SATÔ, M. 1966. *Phytochem.* **5**, 385.
SANNIÉ, C. & SAUVAIN, H. 1952. *Les couleurs des fleurs et des fruits. Anthocyannes et flavones.* Editions du Muséum, Paris.
SAWADA, T. 1958. *J. Pharm. Soc. Japan,* **78**, 1023.
SCARPATI, M. L. & D'AMICO, A. 1960. *Ricerca. Scient.* **30**, 1746.
SCARPATI, M. L. & GUISO, M. 1963. *Annali. Chim.* **53**, 1315.
SCARPATI, M. L. & ORIENTE, G. 1958a. *Tetrahedron Lett.* **4**, 43.
SCARPATI, M. L. & ORIENTE, G. 1958b. *Ricerca Scient.* **28**, 2329.
SCOTT-MONCRIEFF, R. 1936. *J. Genet.* **32**, 117.
SEGUIN, A. 1796. *Annls. Chim. Phys.* **20**, 15.

SEIKEL, M. K. 1962. In: *The chemistry of flavonoid compounds*, ed. by T. A. Geissman. Pergamon Press, Oxford.

SEIKEL, M. K. 1964. In: *Biochemistry of phenolic compounds*, ed. by J. B. Harborne. Academic Press, London & New York.

SEIKEL, M. K., CHOW, J. H. S. & FELDMAN, L. 1966. *Phytochem.* **5**, 439.

SESHADRI, T. R. 1962. In: *The chemistry of flavonoid compounds*. ed. by T. A. Geissman. Pergamon Press, Oxford.

SESHADRI, T. R. 1967. *J. Ind. Chem. Soc.* **44**, 628.

SHAW, B. L. & SIMPSON, T. H. 1955. *J. chem. Soc.* 655.

SHERRATT, H. S. A. 1958. *J. Genet.* **56**, 1.

SIEGELMAN, H. W. 1964. In: *Biochemistry of phenolic compounds*, ed. by J. B. Harborne. Academic Press, London & New York.

SIEGELMAN, H. W. & HENDRICKS, S. B. 1957. *Pl. Physiol., Lancaster*, **32**, 393.

SIEGELMAN, H. W. & HENDRICKS, S. B. 1958. *Pl. Physiol., Lancaster*, **33**, 185.

SLABECKA-SZWEYKOWSKA, A. 1955. *Acta Soc. Bot. Pol.* **24**, 3.

SOMERS, T. C. 1966. *Nature*, **209**, 368.

SONDHEIMER, E. 1958. *Archs. Biochem. Biophys.* **74**, 131.

SONDHEIMER, E. & KERTESZ, Z. I. 1948. *Analyt, Chem.* **20**, 245.

STAFFORD, H. A. 1966. *Pl. Physiol., Lancaster*, **41**, 953.

STAHL, E. & SCHORN, P. J. 1961. *Hoppe Seyler's Z. physiol. Chem.* **325**, 263.

SUMERE VAN, C. E. 1960. In: *Phenolics in plants in health and disease*, ed. by J. B. Pridham. Pergamon Press, Oxford.

SWAIN, T. 1953. *Biochem. J.* **53**, 200.

SWAIN, T. 1954. *Chemy. Ind.* 1144.

SWAIN, T. 1962a. In: *The chemistry of flavonoid compounds*, ed. by T. A. Geissman. Pergamon Press, Oxford.

SWAIN, T. 1962b. In: *Wood extractives*, ed. by W. E. Hillis. Academic Press, London & New York.

SWAIN, T. (ed.) 1963. *Chemical plant taxonomy*. Academic Press, London & New York.

SWAIN, T. 1965a. In: *Chemistry and biochemistry of plant pigments*, ed. by T. W. Goodwin. Academic Press, London & New York.

SWAIN, T. 1965b. In: *Plant biochemistry*, ed. by J. Bonner and J. E. Varner. Academic Press, London & New York.

SWAIN, T. (ed.) 1966. *Comparative phytochemistry*. Academic Press, New York.

SWAIN, T. & BATE-SMITH, E. C. 1962. In *Comparative biochemistry*, Vol. III., ed. by A. M. Florkin and H. S. Mason. Academic Press, New York.

SWAIN, T. & GOLDSTEIN, J. L. 1964. In: *Methods in polyphenol chemistry*, ed. by J. B. Pridham. Pergamon Press, Oxford.
SWAIN, T. & HILLIS, W. E. 1959. *J. Sci. Fd Agric.* **7**, 669.
SWAIN, T. & NORDSTRÖM, C. G. 1957. *Les hétérocycles oxygénés.* Colloquium of C.N.R.S., Paris.
TAYEAU, F., MASQUELIER, J. & LEFÉVRE, G. 1951. *Bull. Soc. pharm. Bordeaux*, **89**, 5.
THIMANN, K. V. & EDMONDSON, Y. H. 1949. *Archs. Biochem.* **22**, 33.
THIMANN, K. V. & RADNER, B. S. 1955a. *Archs. Biochem. Biophys.* **58**, 484.
THIMANN, K. V. & RADNER, B. S. 1955b. *Archs. Biochem. Biophys.* **59**, 511.
THOMAS, M. B. & MABRY, T. J. 1968. *Phytochem.* **7**, 787.
THOMAS, M. B. & MABRY, T. J. 1968. *Tetrahedron Lett.* **24**, 3675.
THOMSON, R. H. 1964. In: *Biochemistry of phenolic compounds*, ed. by J. B. Harborne. Academic Press, London & New York.
TIMBERLAKE, C. F. & BRIDLE, P. 1965. *Chemy. Ind.* 1520.
TOMASZEWSKI, M. 1960. *Bull. Acad. pol. Sci.* **8**, 61.
TOWERS, G. H. N. 1964. In: *Biochemistry of phenolic compounds*, ed. by J. B. Harborne. Academic Press, London & New York.
TRONCHET, J. 1962. *Annls scient. Univ. Besançon*, 2nd series, Bot. **18**, 101.
TRONCHET, J. 1963. *Annls scient. Univ. Besançon*, 2nd series, Bot. **19**, 21 and 25.
TRONCHET, J. 1965. *Annls scient. Univ. Besançon*, 3rd series, Bot. **2**, 12.
UNDERHILL, E. W., WATKIN, J. E. & NEISH, A. C. 1957. *Can. J. Biochem. Physiol.* **35**, 219.
VANCRAENENBROECK, R. & LONTIE, R. 1963. *9th European Brewery Conv., Proc. Cong., Brussels.*
VANCRAENENBROECK, R., ROGIRST, A., LEMAITRE, H. & LONTIE, R. 1963. *Bull. Soc. chim. Belg.* **72**, 619.
VENKATARAMAN, K. 1966. *J. Scient. ind. Res.* **25**, 97.
VILLE, A., COMTE, P., ZWINGELSTEIN, G. & FAVRE-BONVIN, J. 1958. *Bull. Soc. chim.* 1352.
VILLE, A. & PACHÉCO, H. 1967. *Bull. Soc. Chim. biol.* **49**, 657.
VINCE, D. & GRILL, R. 1966. *Photochem. Photobiol.* **5**, 407.
VUATAZ, L., BRANDENBERGER, H. & EGLI, R. H. 1959. *J. of Chromat.* **2**, 173.
WAGNER, H. 1964. In: *Methods in polyphenol chemistry*, ed. by J. B. Pridham. Pergamon Press, Oxford.
WAGNER, H. 1966. In: *Comparative phytochemistry*, ed. by T. Swain. Academic Press, London & New York.
WAISS, A. C. & JURD, L. 1968. *Chemy. Ind.* 742.

WAISS, A. C., LUNDIN, R. & CORSE, J. 1964. *Chemy. Ind.* **28**, 1984.
WARREN STECK. 1968. *Phytochem.* **7**, 1711.
WATANABÉ, S., SAKAMURA, S. & OBATA, Y. 1966. *Agri. biol. Chem.* **30**, 420.
WEINGES, K. & FREUDENBERG, K. 1965. *Chemy Comm.* **11**, 220.
WEINGES, K., KALTENHÄUSER, W., MARX, H. D., NADER, E., NADER, F., PERNER, J. & SEILER, D. 1968. *Justus Liebigs. Annls. Chem.* **711**, 184.
WEINGES, K. & NAGEL, D. 1968. *Phytochem.* **7**, 157.
WERCKMEISTER, P. 1954. *Züchter*, **24**, 224.
WERCKMEISTER, P. 1965. *1st International Symposium on the Iris.* Giuntina, Florence.
WEURMAN, C. & DE ROOIJ, C. 1958. *Chemy. Ind.* 72.
WHALLEY, W. B. 1962. In: *The chemistry of flavonoid compounds*, ed. by T. A. Geissman. Pergamon Press, Oxford.
WHELDALE-ONSLOW, M. 1925. *The anthocyanin pigments of plants.* Cambridge University Press.
WHITE, T. 1956. in *The chemistry of vegetable tannins.* Society of Leather Trades' Chemists, Croydon (England).
WHITE, T. 1957. *J. Sci. Fd Agric.* **7**, 377.
WHITE, T. 1958. In: *Chemistry and technology of leather*, vol. II, p. 111. Reinhold, New York (after Jurd, 1962b).
WILLIAMS, A. H. 1955. *Chemy. Ind.* 120.
WILLIAMS, A. H. 1960. In: *Phenolics in plants in health and disease*, ed. by J. B. Pridham. Pergamon Press, Oxford.
WILLIAMS, A. H. 1966. In: *Comparative phytochemistry*, ed. by T. Swain. Academic Press, London & New York.
WILLIAMS, R. T. 1964. In: *Biochemistry of phenolic compounds*, ed. by J. B. Harborne. Academic Press, London & New York.
WILLSTÄTTER, R. *et al.*, 1915 and 1916. *Justus Liebigs Annln. Chem.*
WILLSTÄTTER, R. & MALLISON, H. 1915. *Justus Liebigs Annln. Chem.* **408**, 147.
WONG, E. 1967. *Phytochem.* **6**, 1227.
WONG, E. 1968. *Phytochem.* **7**, 1751.
WONG, E. & FRANCIS, C. M. 1968. *Phytochem.* **7**, 2123, 2131 and 2139.
WONG, E., MORTIMER, P. P. & GEISSMAN, T. A. 1965. *Phytochem*, 4, 89.
WONG, E. & MOUSTAFA, E. 1966. *Tetrahedron Lett.* **26**, 3021.
ZAPROMETOV, M. N. 1958. *Fiziology Rast.* **5**, 51 (after Neish, 1964).
ZAPROMETOV, M. N. 1959. *Dokl. Akad. Nauk. S.S.S.R.*, **125**, 1359 (after Towers, 1964).
ZAPROMETOV, M. N. 1962. *Biokhimiya*, **27**, 366 (after Neish, 1964).
ZUMAN, P. 1952. *Chemiske Listy*, **46**, 328.

INDEX

3+1 spray
white geese